PROSPECTS FOR DEVELOPMENT IN THE ASIA-PACIFIC AREA

Proceedings of the First Annual
University of Victoria-National Sun Yat-sen University
Social Science Symposium

held at

Dunsmuir Lodge, University of Victoria,
Victoria, BC, Canada

August 12-13, 1999

PROSPECTS FOR DEVELOPMENT IN THE ASIA-PACIFIC AREA

Proceedings of the First Annual
University of Victoria-National Sun Yat-sen University
Social Science Symposium

held at

Dunsmuir Lodge, University of Victoria,
Victoria, BC, Canada

August 12-13, 1999

edited by

Robert E. Bedeski and John A. Schofield

Canadian Western Geographical Series

Volume 37

Western Geographical Press

DEPARTMENT OF GEOGRAPHY, UNIVERSITY OF VICTORIA
P.O. BOX 3050, VICTORIA, BC, CANADA V8W 3P5
PHONE: (250)721-7331 FAX: (250)721-6216
EMAIL: HFOSTER@OFFICE.GEOG.UVIC.CA

Canadian Western Geographical Series

editorial address

Harold D. Foster, Ph.D.
Department of Geography
University of Victoria
Victoria, British Columbia
Canada

Since publication began in 1970 the Western Geographical Series (now the Canadian and the International Western Geographical Series) has been generously supported by the Leon and Thea Koerner Foundation, the Social Science Federation of Canada, the National Centre for Atmospheric Research, the International Geographical Union Congress, the University of Victoria, the Natural Sciences Engineering Research Council of Canada, the Institute of the North American West, the University of Regina, the Potash and Phosphate Institute of Canada, and the Saskatchewan Agriculture and Food Department.

PROSPECTS FOR DEVELOPMENT IN THE ASIA-PACIFIC REGION

University of Victoria-National Sun Yat-sen University Social Science Symposium (1st : 1999 : Victoria, B.C.)

(Canadian western geographical series; 1203-1178; 37)
ISBN 0-919838-27-8

1. Pacific Area–Economic conditions–Congresses. 2. Pacific area–Social conditions. 3. East Asia–Economic conditions–Congresses. 4. East Asia–Social conditions–Congresses. I. Bedeski, Robert E. II. Schofield, John A. III. Series.

HC460.5.U54 1999 330.95 C00-910542-5

Editors' Acknowledgements

We are grateful to the authors and discussants of papers who made the symposium and this volume possible. A list of symposium participants is included at the back of the volume. We are particularly grateful to Professor Peter Lin who not only contributed a paper but worked with us to coordinate arrangements for the symposium. For financial support in helping to host the symposium, we are indebted to Dr. David Strong, President and Vice-Chancellor of the University of Victoria, who has taken the lead in encouraging international initiatives such as our symposium at the University of Victoria. We are also grateful to the Taipei Economic and Cultural Office, Vancouver, for financial support. For assistance with editing we thank Ms. Cecilia Freeman-Ward and Ms. Wendy Goodger.

Technical production was undertaken by the same team that has been responsible for so many of the previous volumes in the Canadian Western Geographical Series: the Technical Services Division of the Department of Geography at the University of Victoria, in particular Mrs. Diane Braithwaite and Mr. Ken Josephson; and Professor Harry Foster, Series Editor.

Contents

SECTION II
Economic Growth and Development in the Asia-Pacific Region

ABBREVIATIONS

ADB	Asian Development Bank
ARATS	Association for Relations Across the Taiwan Straits (PRC)
ASEAN	Association of South-East Asian Nations
CPC	Communist Party of China (PRC)
CTBT	Comprehensive Test Ban Treaty
DDP	Democratic Progressive Party (Taiwan)
DPRK	Democratic People's Republic of Korea
KMT	Kuomintang, or Guomindang (Taiwan)
MAC	Mainland Affairs Council
MFN	Most Favoured Nation
NATO	North Atlantic Treaty Organization
NIE	Newly Industrializing Economies
NPC	National People's Congress (PRC)
PLA	People's Liberation Army
PPP	Purchasing Power Parity
PRC	People's Republic of China
ROC	Republic of China
SEF	Straits Exchange Foundation (Taiwan)
TRA	Taiwan Relations Act (US)
WMD	Weapons of Mass Destruction
WTO	World Trade Organization

List of Figures

List of Tables

List of Plates

Plate 1 A park in commemoration of Dr. Sun Yat-Sen, Kaohsiung

寶島鐘錶
ROLEX
勞力士
OMEGA

Introduction

1

Robert E. Bedeski
Professor, Department of Political Science, University of Victoria

John A. Schofield
Dean, Faculty of Social Sciences, University of Victoria

The chapters in this volume form an eclectic collection around the theme of prospects for development in the Asia-Pacific region. They were presented in their original forms at the University of Victoria, British Columbia, Canada on the occasion of the first annual collaborative symposium of the Faculty of Social Sciences at the University of Victoria and the College of Social Sciences at National Sun Yat-Sen University, Kaohsiung, Taiwan in August 1999. The symposium marked the signing of an academic cooperation agreement between the Faculty and the College of Social Sciences.

While it is impossible in practice to divorce the political and sociological aspects of development from the economic aspects, for convenience of presentation the collection of chapters is divided into two sections along these lines. In the first section we include the chapters that deal primarily with political and sociological issues. Chapters on primarily economic themes are included in the second section.

Section One: State and Society in Taiwan and Canada

Tensions in the Taiwan Straits

Several chapters reflect on tensions in the Taiwan Straits between the People's Republic of China (PRC) and the Republic of China (ROC). In 1949, the ROC had established its base on Taiwan, and proceeded to demonstrate a significant model of economic development and industrialization in the face of major difficulties, and then from the early 1980s, to move to full multi-party democracy. Taipei has skilfully navigated its security and diplomacy in the face of military threats from Beijing, with varying degrees of support from the US.

Gerard S.H. Chow examines the Taipei–Washington–Beijing triangle, in which Taiwan consolidates its position and makes the triangular relations not only possible, but effective. He analyses President Lee's 1999 statement on "special state-to-state relations," and examines reactions from the US and the PRC. His chapter also explores the possibility of future talks between Taiwan and China. Taiwan must involve other states if it is to survive, and so must avoid reduction to simple bilateral relations. The PRC, on the other hand, proclaims the Taiwan issue as one between itself and a secessionist province, brooking no "interference" from the rest of the world.

Marion Wang analyses Taiwan's diplomacy as the ROC searches to restore its international status. Past and present contributions to the UN and its programs justify her claim to membership, but PRC pressure on other member states continues to exclude it. Taiwan wants representation for her 22 million citizens in international organizations, and not independence. Beijing has sought to reduce the number of ROC allies to zero by the year 2000, but has not been successful. Nevertheless progress has been made in Cross-Strait dialogue, some democratization on the mainland, and reduction of hostilities.

Robert Bedeski examines arms control and disarmament in Asia, with attention to three regions of potential conflict—the Korean Peninsula, the Taiwan Straits, and India-Pakistan. Modern historical experience seems to demonstrate that security deficits have contributed to conflicts on the Asian continent. Furthermore, even disarmament at the social level does not ensure peace and security. As long as the Democratic People's Republic of Korea (DPRK) perceives a security deficit vis-à-vis the US, it will be attracted to a weapons of mass destruction (WMD) solution. The ROC, on the other hand, will rely more on alliances than a nuclear program. Most disturbing in recent years is the nuclear testing in South Asia, and India's justifications for it.

Social Change in Taiwan

As one of the most successful modernizers in the modern world, Taiwan has been undergoing some major changes. It has seen waves of migration from the mainland—most recently after the defeat of the Guomindang following the 1949 Communist revolution. Jou-juo Chu examines aboriginal peoples on the island who are organized into a number of tribes that have territory dispersed in different types of topography. As living standards have improved and wage levels increased, the aboriginals have been a source of cheaper labour for many small enterprises. Many have adapted to the urban and industrial environment, but very often they are exploited by operators. With the labour shortage, employers have sought to import foreign labourers, which negatively affects the bargaining power of aboriginals and others just entering the work force.

Using the sociological framework of organizational theory and empirical research, Stanley T. Lee studies the role of leadership in organizational commitment in a Taiwan steel company. He analyses data from 1,299 questionnaires, and hypothesizes that leader ability and attitude have a significant impact on organizational commitment with respect to work units and other organizational structures, and also through the work problems perceived and experienced by subordinates. Work problems were found to be negatively associated with organizational commitment, so when people are committed to their group, they gladly accept whatever task or work role is assigned to them.

Canada and the Asia Pacific

While Canada and Taiwan may be as far apart as any two nations on earth, there is increasing linkage, or at least emerging parallel concerns and interests. Globalization is one of these concerns, and once isolated communities are becoming part of the conflicts generated by change. Warren Magnusson focuses on the conflict over Clayoquot Sound, where a trans-national forest company was confronted by a coalition of citizens blocking the clear cutting of primeval forest. Drawing from that struggle, he examines three issues related to the globalization of political space: (1) the articulation of "the market" as a political space; (2) the proliferation of "identities" in political struggle; and (3) the transfiguration of sovereignty as a disciplinary discourse. He sees an increasing politicization of the market and the emergence of a "consumerate" reflected in the sophisticated campaign against the corporate structures.

The increasing immigration from Asia has been a growing political issue in Canada. Zheng Wu (with Violet Kaspar) analyses survey data to show some of the stresses of migration, and how resilient Asians have been in dealing with them. Using the stress process model, they evaluate three hypotheses on mental health differences. Underlying their analysis is the view that voluntary immigrants share a "so-called 'hardy' personality," enabling them to better cope with stresses of a new cultural, political, and economic environment. While they allow that culture may inhibit many Asians from seeking psychological help with distress, their tentative conclusion is that the immigrants may be better prepared for the new stresses.

Section Two: Economic Growth and Development in the Asia-Pacific Region

Economic Interdependence

Carl Mosk looks at the processes of convergence and divergence in such measures as standards of living, knowledge of technology, and per capita levels of human and physical capital as economic growth proceeds. The net effect of various forces, including population migration, is to strengthen convergence, and this in turn has promoted increasing integration of the economies of the region. Specifically, he shows how the trend toward convergence dominates the process of economic adjustment under way in the two economies of Japan and British Columbia (BC). He also demonstrates how migration has shaped the process of adjustment in these and other Asia-Pacific economies. As he puts it, 'convergence begets demographic integration which in turn contributes to convergence.' However, using Japan and BC as examples, he concludes that it is not possible to say whether any two countries—or regions—will become more or less integrated on a bilateral basis. Nonetheless, in the long-run, convergence in the Asian sub-region of the Asia-Pacific will re-emerge as an engine of growth in BC.

Although the Mosk chapter does not extend to an analysis of the relationship between Taiwan and mainland China, it leaves open the implication that the two economies could converge, producing greater integration, further convergence and, possibly, less tension over the 'Taiwan issue.' These are matters taken up in detail in the Liu and Keng chapters.

Tru-Gin Liu adopts a game-theoretic approach to studying the effects of recent changes in the relationship between China, Taiwan, and the US on Taiwan's trade policy toward China. The model predicts that Taiwan's best strategy is to use negotiations between the US and China over the 'Taiwan issue' to serve its own political goal of independence. Rather than necessarily continuing its patient Mercantilist policy toward China, the message is that Taiwan should maintain a flexible approach in trading across the Taiwan Straits because it has to take into account the impact on the US of its economic links with China.

Kenneth Keng reviews long-term growth prospects in China to 2020, and emphasizes that the key constraint on economic growth will be the shortage of plains land for transportation and urbanization. The limited supply of plains land will compel China to develop relatively independent regional economies so that long-distance transportation needs are minimized. Professor Keng identifies nine such regions, each comprising a network of urban centres made up of one or two metropolitan areas and a number of smaller satellite cities with rural areas in between. He goes on to predict that enhanced growth through this regionalization process will accelerate integration of Hong Kong (with Macao) and Taiwan with their neighbouring mainland economies of Guangdong and Fujian to give still greater economic potential to the integrated regions. Since Hong Kong and Taiwan are likely to provide most of the capital resources required to fuel growth in, respectively, Guangdong and Fujian, China may be able to direct its scarce capital resources to the development of other mainland regions. It is therefore possible, as he sees it, for Greater Shanghai and the South-Central region to join Taiwan-Fujian and Hong Kong-Guangdong as the new 'Asian Tigers' of the 21st century.

Labour Force Participation

Noting that the level of labour force participation by women, and its growth since 1951, have been lower in Taiwan than in industrialized countries, Peter Lin investigates various aspects of female participation in order to get a deeper understanding of its causes and effects. Female labour force participation clearly influences the aggregate supply of labour and hence the capacity of the economy for growth and development. Using a logit model, he estimates participation rates for women with different characteristics. Participation is highest for women who are unmarried, 36-45 years of age, have a university education and prior work experience. Everything else being equal, residence in rural areas gives a higher probability of labour force participation than residence in urban areas, and living with in-laws also boosts the participation rate, as does household income. The effect on labour supply is then estimated over the period 1999 through 2003 of increasing the labour force participation rate by 0.1% and 0.2% each quarter during that period. Finally, using factor analysis, the relative importance of changes in population structure and willingness to participate in the labour force is estimated for the period 1980 to 1997 as each affects the female participation rate. The effect of changes in willingness to participate is most marked in the 25-49 age group. From these several findings, inferences can be drawn for policy-making in respect of female participation in the workforce.

Theory and Practice

Chingnun Lee tests the validity of the purchasing power parity (PPP) theory with respect to exchange rates between the currencies of Taiwan and selected major industrial countries. The PPP theory indicates that the nominal exchange rate should equal the ratio of prices in any two countries. Using unit root tests and fractional co-integration methods, Professor Lee finds that the theory applies in terms of exchange rates in Taiwan. While the adjustment process toward the equilibrium exchange rate based on PPP is long-term in the cases of the US and the UK, it is short-term in the cases of Germany, France, and Canada. As another research project, it should be instructive to explain the implications of these differences for the economic progress of Taiwan.

In the final chapter in the volume, Don Hong presents another test of theory against practice. He outlines the simple theory of monetary policy, showing through various transmission mechanisms that increases in the money supply or decreases in the interest rate increase national income. Using recent Japanese data, he then shows that the theoretically inverse relationship between the money supply and short-term interest rates does not always hold in practice. He also makes the argument that even though manipulation of the money supply is theoretically the primary means of influencing economic activity through monetary policy, both Japanese and US authorities have preferred to use interest rates as their main policy instrument in recent times. An implication of this strategy to which he refers is that if the Japanese economy does not quickly recover from recession and US interest rates continue to rise, trade tensions could develop between the two countries leading to increased protectionism. Presumably, a further implication could be stalled economic growth in the Asia-Pacific region, as well as elsewhere.

Plate 2 The Gods of Sheng Huang Temple, Kaohsiung ➧

Section 1

State and Society in Taiwan and Canada

2

Cross-Strait Relations in 1999: Political Dialogue or Negotiation? From "One China" to "Special State-to-State Relations"

Gerard S. H. Chow

Director, Institute of Mainland Studies, National Sun Yat-Sen University

Introduction

It is common sense that the sum of two sides of a triangle is always larger than the remaining side in a triangular diagram. Although Taiwan (ROC), in terms of its size, population, military force, etc. is a relatively small state compared to the PRC and the US, it must bring Beijing and Washington into a real triangular diagram. The real triangular structure must be defined as the three actors interacting substantially and mutually, and formulating a genuine triangular relationship where the actors can influence each other.

For a small state like Taiwan, the strategy of dealing with the PRC is to invite international participation from the rest of world. In bilateral relations between Taiwan and China, assuming that there are no foreign involvements, Taiwan would face a different situation when talking about cross-strait relations. Therefore, the strategy for Taiwan must be to avoid the bilateral "China-Taiwan" formula, as well as a loose Taiwan-China-US triangle.

In a genuine triangular relationship (Figure 2.1, Diagram One), Taiwan is able to justify its argument with Washington's endorsement, then the mathematical axiom mentioned above would favour Taiwan. On the other hand, in a loose triangle, or one based on bilateral relations (Figure 2.1, Diagram Two), Taiwan may suffer pressures from its opponent, Beijing, although there might be some voice from Washington indirectly and passively.

This chapter will examine how Taiwan consolidates its position and makes the Washington-Taipei-Beijing triangle work well. By analysing President Lee's statement on "special state-to-state relations," the chapter examines reactions from the US and the PRC, as well as exploring the possibility of fruitful talks between Taiwan and China in the near future. Based on President Lee's statement, this chapter will discuss the latest scenarios that have occurred in Washington-Taipei-Beijing relations.

◀ **Plate 3** Chiang Kai-shek Memorial, Taipei

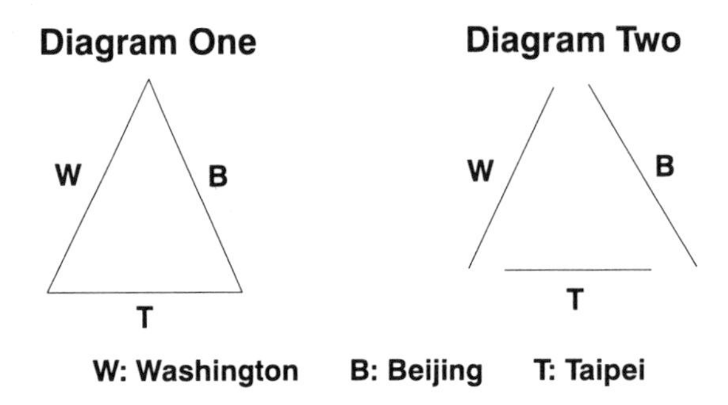

Figure 2.1 Trilateral relations

TAIWAN'S INITIATIVE

During an interview with the *Voice of Germany* (*Deutsche Welle*) on July 9 1999, President Lee Teng-hui expressed for the first time that the cross-strait relationship had evolved into a state-to-state relationship, or at least into "special state-to-state relationships," and should not be viewed under the one China framework. The one China policy can be defined as an internal relationship between a legitimate government and a rebellious province or a central government and a local government. The "state-to-state" relationship clearly has surpassed the "two political entities" definition promulgated in the National Unification Guidelines and rejects the Beijing overture of "one country, two systems" model toward Taiwan.[1]

President Lee's statement was based on history and legal aspects. The historical facts show that the PRC government has never governed the ROC's territories since they were founded in 1949. Legally speaking, the ROC's government has strictly confined its territorial jurisdiction to Taiwan since 1991 when the National Assembly completed the constitutional reforms. Due to the 1991 reform, the PRC was recognized as a legitimate country, ruling over the mainland of China. Furthermore, the constitutional reforms in 1992 stated that: (1) the president and vice president of ROC will be elected directly from the Taiwanese people; and (2) the government in Taiwan only represents the people of Taiwan, in which the legitimacy of government's ruling power has nothing to do with the Chinese people.[2]

The Chinese government never seriously considered that a divided China or two different states existed between Mainland China and Taiwan, and so Beijing has continued to intimidate Taiwan by using force. These are the main reasons why the cross-strait relations had not improved.

Following President Lee's action, Su Chi, chairman of the Mainland Affairs Council (MAC), reiterated at a news conference, "We have shown our good will by calling ourselves a political entity under one China policy, but the Chinese communists have continued to use this policy to squeeze us internationally.

We feel there is no need to continue using the 'one China' term. China can't continue to put the 'one China' hat on Taiwan."[3] Su Chi stressed that the practical realities of the situation are that China has never controlled Taiwan and its adjoining islands, Kinmen, Matsu, and Penghu, and Taiwan has its own constitution as well as elected officials. Su pointed out, however, that because of Taiwan and China's ethnic and cultural ties, the relationship between China and Taiwan could be closer than relations between most states. The relationship between the two could be "one nation, two states"—similar to the German unification model. He explained the cross-strait relations should be "state-to-state" or "interstate relations."[4]

The other important statements from Su Chi are that the new policy will have no immediate or major impact on Taiwan's mainland policy. "A peaceful reunification with a democratic China remains the ultimate goal underlying the president's new policy." Vice President Lien Chan clearly said the government has not changed its mainland policy.[5] Su Chi urged Wang Dao-han to pay a visit in the fall. In the next meeting between Koo Chen-fu and Wang Dao-han, there could be political talks in which the agenda may be arranged from technical issues to political issues, such as state sovereignty, and national unification.

President Lee had to be aware that his statement would cause a stir domestically and internationally. In the context of recent strains in the US-Sino relationship, China and the US were shocked by the President's "special state-to-state" policy announcement. Why did he make remarks that went beyond what officials in Taiwan have said before? And why now?

Mainstream speculation has been that the president revived the touchy subject of the status of the island vis-à-vis the mainland for domestic political reasons. However, there is more at stake for Taiwan than the question of whose candidate wins the 2000 presidential elections. As important as positioning the KMT (Kuomintang, or Nationalist Party) for the election might be, President Lee may have been trying to position the US. He knows very well that Washington's attitude toward Taiwan, and specifically the credibility of a US commitment to defend the principle of peaceful reunification, is the key to the island's future.[6]

There are at least two explanations of "why now?" As noted above, the first reason is the strain in the US-China relationship. Relations between the two giants are at their lowest point in recent years. China's reported espionage campaign to secure America's nuclear secrets, its illegal efforts to obtain US ballistic missile technology, allegations of unlawful campaign contributions, disputes over most favoured nation (MFN) status, and its application to the World Trade Organization (WTO), have all made it politically tough to generate sympathy in Washington for Beijing's position. Besides these disputes, the mistake of the North Atlantic Treaty Organization (NATO) in bombing the Chinese embassy in Belgrade serves as the key reason for the crises between the two countries.

By Taiwan's calculation, China's military capability is substantially smaller than what it might be in coming years if the People's Liberation Army (PLA) takes advantage of nuclear and missile technologies. President Lee may have concluded that shifting the status of cross-strait relations today may, in fact, be less costly than trying to create new status under the next president.[7] One thing is certain: among the four presidential candidates (Lien, Sung, Chen, and Hsu), no one will hold as strong a position as President Lee in Taiwan.

What has been the reaction of the Taiwanese people toward President Lee's "special state-to-state relations" model? A survey conducted by the Chinese Association for EuroAsia Studies, a private organization in Taipei, found that 73.3% of 1,103 people polled agreed with Lee's statement, while only 17.3% disagreed.[8] Lee's redefinition of cross-strait relations caused anger in Beijing, but internally received strong support from the public.

Beijing's Response

Immediately after President Lee's statement, the Chinese government reacted strongly. The Foreign Ministry spokesman Zhu Bangzao said on state television, "We can clearly see that he [Lee Teng-hui] has gone a long way down the road of playing with fire. We sternly warn Lee Teng-hui and the Taiwan authorities not to underestimate the Chinese government's firm determination to uphold the nation's sovereignty, dignity, and territorial integrity."[9] Zhu continued, "Don't underestimate the courage and force of the Chinese people to oppose separatism and Taiwan independence. China's reunification is the trend of times and the will of the Chinese people."[10] Zhu finally concluded on a threatening note, "Lee Teng-hui and the Taiwan authorities must recognize the situation, rein in at the brink of the precipice, and immediately stop all activities aimed at splitting the motherland."[11]

The chairman of the Association for Relations Across the Taiwan Straits (ARATS), Wang Daohan, said he was "surprised" by Lee's statement that talks between Taipei and Beijing were based on statehood.[12] Wang had hinted that his visit to Taiwan in the fall was now in doubt. Wang sought clarification from Koo Chen-fu, chairman of Strait Exchange Foundation (SEF), who supported Lee Teng-hui, arguing cross-strait talks are state-to-state. Wang even pointed out "the basis for contacts, exchanges and talks between ARATS and SEF no longer existed."[13] In Hong Kong (July 12), ARATS vice chairman Tang Shubei urged that both sides go back to the consensus reached in 1993, which allowed both sides to freely express their definition of "one China."

On July 15, Beijing announced it had successfully developed the design technology to make neutron bombs and miniaturized nuclear weapons on its own. It was the first time that Beijing had made these assertions publicly. The announcement came just a few days after Taipei restated its stance toward

China. Most analysts believe that the timing of the announcement follows China's long-standing policy that Beijing always had the right to use military force to achieve its cherished goal of reunification with Taiwan.[14] Earlier than this announcement, Chi Hao-tien, the Chinese Defence Minister, pointed out that the PLA has been ready to defend territorial integrity, and to crush any separatists.[15]

In view of Taiwan's new statement over cross-strait relations, Beijing reacted by rejecting Lee Teng-hui. In an official press conference, a Japanese reporter asked a question, referring to Taiwanese leader Lee Teng-hui by his official title "President." Zhang Qiyue, Chinese Foreign Ministry spokesman, said before responding, "I would like to correct the title you mentioned. We don't recognize the so-called 'president,' and in the future you should pay attention to this."[16] Zhang's comment highlights the tension between China and Taiwan, and indicates that officials in Beijing seize any opportunity to deny Taiwan and the ROC as a legitimate and sovereign state.

Koo Chen-fu, Taiwan's top negotiator, waited for 3 weeks after July 9 before reiterating an official statement in response to the earlier pledge of Wang Dao-han. The main themes of Koo's statement struck the target. Koo stressed that, "...President Lee's remarks clearly spelled out the reality that ROC is a sovereign state, that the two sides of the Taiwan Strait are ruled separately on an equal basis. The special state-to-state relationships are primarily to lay a foundation of parity between the two sides for promoting the level of dialogue ...In 1992, one China with each side being entitled to its own interpretation was reached at a verbal consensus. What we see as 'one China' is something for the future since China nowadays is divided and ruled separately by two sovereign states. Chairman Jiang Ze-min also has stated that so long as both sides engage in negotiations based on parity, a mutually acceptable resolution will eventually be found."[17]

Again, Beijing reacted angrily. Zhang Ming-qing, the spokesman for the Taiwan Affairs Office under the State Council, said that Koo's statements were not different from Lee Teng-hui's. In reality, his statement is the same as Lee's two-state theory.[18] Zhang remarked that Lee Teng-hui's two-state theory strips away the foundation for contacts, communications, and talks between both sides. Moreover, the ARATS officials immediately dismissed Koo's clarification, saying it had broken pledges to uphold a one China policy, which was a grave violation of the 1992 agreement to uphold mutual recognition of the one China principle. The ARATS was expected to come up with a more forceful statement in the days to come. Koo's open statement was sent to Beijing in a letter, and was bluntly rejected and returned unopened.[19] Beijing required a statement from Koo that would support the one China policy and dismantle the two state theorem. Beijing would then step down from its high position. Unfortunately, Koo's statement did not satisfy Beijing's much-awaited expectation. At present (late 1999), the cross-strait relations seem to be in a deadlock.

Washington's "Intervention"

Washington responded to Taiwan's initiative with caution and clarity. In a US State Department news briefing on July 13, one reporter posed a question to spokesman James Rubin: "Would the US continue to insist that there is only 'one China' if the Chinese themselves say there isn't one China any more?" Rubin avoided a direct answer, instead reiterating the "three no's policy": (a) we do not support Taiwan independence; (b) we do not support Taiwanese membership in organizations where statehood is required; (c) we do not support a two-China policy or a one China/one Taiwan policy.[20] However, after the 40-minute meeting between President Lee Teng-hui and Darryl Johnson, the Director of the American Institute in Taiwan's Taipei office, there was a news briefing held on July 14. From this news release, Washington's position was clear.

The spokesman stressed on behalf of the US government that both President Lee's initial statement and China's reaction to these statements were not beneficial to creating the necessary cross-strait dialogue. However, the official Taiwanese version of the meeting between President Lee and Darryl Johnson indicated that there has been no change in policy that will yield the prospect of a continuation of the cross-strait dialogue. The only realistic way to achieve progress between China and Taiwan is to have a cross-strait dialogue with meaningful, substantive exchanges. The main theme of Rubin's statement was to continue the dialogue, despite the fact that Beijing accused President Lee of playing with fire.

Washington continued to watch and express its concerns over the interactions across the Taiwan Straits. In the US State Department news conference, a reporter asked "How did the US government respond when China revealed its neutron bomb capability and brandished this weapon in a rhetorical way?" Rubin answered "...as far as I know, this was a response to the Cox Committee Report, and we don't see any connection to the Taiwan incident."[21] Whether the Chinese government's action of showing its neutron bomb publicly was on purpose or not, the feeling of Taiwanese people was very strongly against Chinese hegemony over the island. However, a statement from the Washington spokesman said the US government intended to diminish the potential conflict between Taipei and Beijing, thus not recognizing the Chinese nuclear threats toward Taiwan.

The most sensitive question would be if Taiwan were to declare independence in a way that brought about a conflict with China. Would the US still feel it has an obligation to defend Taiwan? Washington responded cautiously and ambiguously. James Rubin said,

> *...as far as our position on China's potential use of force against Taiwan, we have said the future of Taiwan is a matter of the Chinese people. We have an abiding interest that any solution be peaceful.*

> *We encourage the two sides to engage in a substantive, cross-strait dialogue. We also, in the Taiwan Relations Act, provide that it's our policy to consider any effort to determine the future of Taiwan by other than peaceful means. A threat to the peace and security of the Western Pacific is of grave concern to the United States."*[22]

Rubin came to the conclusion that "I don't think any of the statements that make cross-strait dialogue less likely or less likely to be successful are helpful. I don't think any reactions to statements that are less likely to yield a dialogue or a successful dialogue are helpful." This is the first time that the US official statement put its fundamental concepts together. The key concepts of the US government with regard to the crisis of Taiwan Straits are dialogue, peaceful means, and the Taiwan Relations Act. With the three main components of the US policy toward China and Taiwan, Washington was able to find a way to cast its influence on the Chinese affairs, and seemed to "intervene" legally and legitimately.

Cross-strait relations became tense when Beijing reasserted military force to counter Taiwan's independence. President Clinton telephoned President Jiang Zemin on July 19, 1999 assuring Beijing that the long-standing position of the US continues to be that there is one China, as well as that the US has an abiding interest that any resolution be peaceful. The two sides are encouraged to engage in meaningful, substantive dialogue, and Clinton also made it plain that deterioration of the atmosphere of the Strait is in no one's interest. Again, it is quite clear that instead of ambiguities as the US provided in the past, the conversations between Clinton and Jiang indicated that Washington had moved toward an active involvement in triangular relations. What Clinton told Jiang might not be different from the past, but the implications from the phone call were crucial to dilute the crisis. Analysts even point out that Washington's active and direct involvement in mediating curbed Beijing's reaction, though Washington for a long time has claimed that it will never act as an "arbitrator" or "mediator" between China and Taiwan. The US has thus become a significant actor in the cross-strait issue without directly seeking to mediate relations between the two sides. Both sides regard the American posture as potentially of critical significance in structuring both the psychological and the actual military balance across the Strait. In effect, this is a three-sided issue, even though Beijing regards it as strictly a domestic matter.[23]

On July 20, 1999, President Clinton made a statement later released by the Office of the Press Secretary. Clinton said he had a conversation with Jiang Ze-min in which he restated strong support of the one China policy as well as for the cross-strait dialogue. He also made it clear that the US policy had not changed, including the view under the Taiwan Relations Act. The US would take very seriously any abridgement of the peaceful dialogue.[24] From this statement, dialogue, peaceful solutions, the Taiwan Relations Act, and perhaps the one China policy, come to the fore. The President made it clear that he believes the best way to resolve any issue between the two sides is in the context of the cross-strait dialogue.[25] During the press conference, the US government was asked the question "Is it our policy to use force to protect Taiwan if attacked?" The government responded, "...our policy is outlined in Taiwan's Relations Act, and that has not changed. ...we have commitments as far as Taiwan's self-defence, and it's spelled out clearly in Taiwan's Relations Act..." The question continued "Is the US one China policy consistent with state-to-state dialogue between China and Taiwan?" The White House answered "...our policy has not changed as far as one China, and in our conversations with Taiwan (Darryl Johnson's meeting with President Lee), they have indicated their policy has not changed."[26] From this statement, Washington has successfully played its "balancing role" in which Taipei and Beijing have to listen to the voice from America, and perhaps more importantly that the three actors link tightly, thus forming a triangular structure. It should be noted that the pushing force to construct the triangle was from Taiwan, particularly the invisible hands of President Lee Teng-hui.

Washington did not just wait and see. President Clinton engaged in the Chinese affairs with action. In order to manage the changing situation and understand what Beijing and Taipei are thinking, he sent two special envoys to Taiwan and China.

Richard Bush, the Chairman of the American Institute in Taiwan, flew to Taipei on July 22 for discussions with the Taiwanese government on a wide range of issues, including cross-strait matters and Taiwan's cross-strait policy.[27] Meanwhile, Stanley Roth and Kenneth Lieberthal arrived in Beijing for consultation with Beijing's top leaders. These actions produced the following results: (a) calming down the tension in the cross-strait relations; (b) restating the US policy toward Beijing and Taipei; and (c) reaffirming Washington's determination to engage in China/Taiwan debate.

Tokyo, echoing the US position, said "one China" is ruled from Beijing and expressed hopes that Beijing and Taipei would work out their problems peacefully. Sadaaki Numata, spokesman for the Foreign Ministry said, "There is no change at all in the Japanese government's position on Taiwan," and expressed that "Tokyo urged the two sides to settle their disputes by talking."[28]

One point should be highlighted. Before President Lee opened the new issue, the US was putting pressure on both sides of the Taiwan Strait to reach some form of so-called "interim agreements" in scheduled talks between the top negotiators of both sides this fall. In a stronger statement than ever before, Stanley Roth, the Assistant Secretary of State for East Asian and Pacific Affairs, said on June 29 that "mere technical agreements about some of the minor aspects of cross-straits relations...is not going to do a whole heck of a lot to improve the relationship." "Interim agreement would be able to address significant issues, more than technical issues and less than a comprehensive resolution," Roth continued. However, the US suggestions seem to have come to a halt, due to the upcoming situation.

Dialogue or Nothing

Since 1995 there has been no dialogue. Recent cross-strait exchanges have amounted to little more than talks about the possibility of meeting to talk. In the meantime, adhering to the "one China" concept has certainly worked to Taiwan's disadvantage. Both Taiwanese and foreigners have sometimes pointed out Taiwan's "one country, two political entities" idea was working in the subtleties across the Taiwan Straits, though Beijing never accepted this concept. To the Chinese government "one China" means the People's Republic of China, and Taiwan's adherence of this concept had seemed to go along with China's portrayal of a renegade province.[29] The PRC government argued that Beijing had not insisted that cross-strait negotiations be conducted under the names of central government and local government. In any case, cross-strait talks should be based on five principles:[30]

1. Both sides should agree to the "one China principle." The talks should not be based on "two countries," or "one China, one Taiwan." It is entirely a matter of Chinese internal affairs, so there is no concession at all.
2. Both sides are equal. Beijing did not insist that cross-strait talks be conducted under the name of central government vs. local government. Both sides could negotiate and discuss negotiation procedure under whatever titles.
3. There would be no constraints regarding the negotiation agenda, only that both sides honour the "one China principle."
4. No foreign power should be allowed to settle cross-strait problems.
5. Under authorization of both governments, the ARATS and SEF could discuss negotiation procedure regarding the agenda and other political issues.

The problem is that there were at least three sets of definitions of "one China." Beijing, Washington, and Taipei all have their definitions of the two words. China and Taiwan agreed to disagree on the definition of "one China" in 1993 to side step the issue of sovereignty. The US keeps the principle of recognizing only one of two governments as the legal government of China.[31] Under such ambiguous definitions, the three-sided balance was maintained. Beijing redefining one China "as the PRC, while Taiwan being a part of its territory" is regarded as an unacceptable position to the Taiwanese government.

What is the definition of "one China" according to the US government? President Clinton announced a "three no's policy" last year when he visited China. He stated that the US would not support Taiwan independence, two Chinas, one China and one Taiwan, and would not support Taiwan joining international organizations that require statehood.[32] Washington obviously did not change its position on the "one China" policy, even though President Lee Teng-hui has moved from "one China, two political entities" to "special state-to-state."

On December 4, 1997, the *People's Daily* issued an article presenting "one country, two systems as the best option for motherland's reunification." This article strongly urged two principles: (a) insistence that the one China principle is the condition and foundation for peaceful negotiation; and (b) insistence that two systems' coexistence provides opportunity for cross-strait talks.[33] It is obvious that the Beijing government was selling the concept of one China to the Taiwanese people. The ultimate target is to keep the principle of one China at hand, while using a flexible strategy—two systems as its mouse trapping.

Furthermore, on December 10, 1997, the Beijing-based journal *Cross-Strait Relations* issued another article, urging that it is inevitable to reopen the cross-strait talks. Again, Beijing did not stray from its firm position. In this article, Beijing proposed a solution: under the one China principle, the two sides end up in a state of hostility. For the sake of termination of cross-strait animosity, the ARATS and the SEF should prepare negotiation for political talks. Beijing's policy was put forward in this article, saying the major hurdle for the talks of technical issues in the past was purely a political question. If the two sides reopen political talks for political issues, or open negotiation procedures, it will be a good opportunity to reopen economic and technical talks for both sides.[34]

The Chinese policy was further outlined on December 14, 1997 in Hong Kong's *Wen Wei Po*, the pro-Beijing newspaper, in an article which strongly attacked Taiwan. The newspaper highlighted: (a) it is unrealistic for the two sides to hold talks on economic and technical issues; and (b) if Taiwan is serious on cross-strait relations, the government should consider the termination of hostility as a priority. For this reason, talks on political issues, such as consultation procedures, should be arranged prior to technical issues as Taiwan proposed to do.[35]

China has continued to insist that Taiwan give up its sovereignty and enter into political negotiation, while Taiwan, recognizing that the issue of sovereignty raises only conflicts and disagreements, and may well become an obstacle to resolving other urgent and practical issues, has opted for talks on fundamental issues.

The facts are that the ROC in Taiwan and the PRC have been two separate sovereign states for the past 50 years and that the area of Taiwan has never been under the jurisdiction of the PRC. Why does Beijing deny two states or two political entities? Taiwan argues that even UN members have shown that "statehood" often follows a far more pragmatic course than one might expect from a purely legal analysis. India was a founding member state of the UN, even though it was still part of the British Empire. The Ukrainian and Byelorussian Soviet Socialist Republics were founding members even though the USSR was also a founding member. Through the decades, the UN has seen membership by the two Germanies, the two Yemens, and the two Koreas. There is no reason to deny that Taiwan exists as a real state. Any suppression of Taiwan's statehood may cause hostility in the Taiwanese people.

Since 1993, Taiwan has tried such formulas as "one China-oriented transitional two-China policy," "one China, but not right now," "one divided China," "one China, two political entities," as well as "one China, two governments."

All seemed ambiguous to outsiders, but they were used to describe the reality: there are two Chinas.[36] However, Taiwan has already taken a step backward by saying it is willing to pursue parallel talks, where political and economic issues can be discussed at the same time.

China pressures Taiwan to come to the political negotiation talks to discuss procedural issues and hold cross-strait forums, using Taiwan's hesitancy to foster an impression of the Taipei government as frigid and obstinate.[37] Taiwan may face increasingly severe political, economic, and military pressure from China. But accepting "one country, two systems," or "one China's PRC" for the sake of peace will go against the will of 80% of the Taiwanese people who are showing their strong attitude against the formulation of "Taiwan is a part of China."

Therefore, Beijing broke off talks with Taipei in 1995 using the excuse of President Lee Teng-hui's visit to the US. From that moment, three agreements were left in limbo—the repatriation of illegal immigrants, the handling of hijackers, and the resolution of legal disputes. "How can we resolve political issues if we can't solve small problems?" said Wu Hsin-hsing, Deputy Secretary General of the Straits Exchange Foundation. China is using the unfinished 1995 agreements as bargaining chips. Bau Tzong-ho, Chairman of the Political Science Department at National Taiwan University, said Beijing is unwilling to talk about economics before politics.[38]

Concluding Notes

The "one China" myth is apparently no longer useful in dealing with cross-strait relations. It is certainly unrealistic to expect Taiwan to proceed with negotiations on an agenda that was based on a myth. The tacit agreement of "one China" had in the past been defined unilaterally by China. A myth sets no base for mutual trust and confidence.

For bilateral talks to proceed in a constructive manner, both sides must take part on an equal footing. China has continued to insist that Taiwan must give up its sovereignty and enter into political negotiations, and has assured that as long as Taiwan follows the "one China principle" everything is negotiable, while Taiwan, recognizing that the issue of sovereignty raises only disagreement and may well become a stumbling block to resolving other more urgent issues, has opted for talks on technical and functional issues.

From July, 1999, according to the President's new guidelines, Taiwan is willing to engage in political talks as long as Beijing recognizes the ROC as a sovereign state. Taiwan has kicked the ball into China's court. More importantly, the jury is also ready. Beijing must find a way of dealing with Taiwan and Washington, but the question is: besides declaring war, does China have any way out of this dilemma other than using peaceful means such as talks or negotiation? It tests Chinese wisdom.

Endnotes

1 http://wep.oop.gov.tw/OOPSRCH.NSF/

2 Ibid.

3 *Taipei Times*, July 13, 1999,p.1.

4 Ibid.

5 *Taiwan News*, July 13, 1999, p.1.

6 *The Asian Wall Street Journal*, July 15, 1999.

7 John Bolton, A Declaration of Independence for Taiwan, *Taipei Times*, July 19, 1999, p.9.

8 Van Tran, Survey shows Lee has wide support, *Taiwan News*, July 19, 1999.

9 *South China Morning Post*, July 13, 1999.

10 *People's Daily*, July 13, 1999.

11 *People's Daily*, July 13, 1999.

12 *Xinhua News Agency*, July 12, 1999, *People's Daily*, July 13, 1999, p.1.

13 *Wen Wei Po*, July 13, 1999.

14 *People's Daily*, July 16, 1999.

15 *Ta Kung Pao*, July 15, 1999.

16 *Taiwan News*, July 21, 1999.

17 *The Liberty Times*, Taipei, July 31, 1999, p.2. *Taiwan News*, Taipei, July 31, 1999, p.2.

18 *Taipei Times*, July 31, 1999, p.1.

19 Ibid, *Taiwan News*, July 31, 1999.

20 US Department of State, *Daily Press Briefing*, July 13, 1999.

21 US Department of State, *Daily Press Briefing*, July 15, 1999.

22 Ibid.

23 Kenneth Lieberthal, "Cross-Strait Relations," paper delivered at the conference on the PRC after the Fifteenth Party Congress—Reassessing the Post-Deng Political and Economic Prospects, Taipei, Institute for National Policy Research, Feb. 19-20, 1998.

24 White House Briefing, statement by the President William J. Clinton, July 20, 1999.

25 White House Briefing, Press briefing by Joe Lockhart, July 20, 1999.

26 Ibid.

27 US Department of State, *Daily Press Briefing*, July 21, 1999.

[28] *Taipei Times*, July 14, 1999.

[29] *Taipei Times*, July 13, 1999, p.8.

[30] *Wen Wei Po*, July 28, 1997.

[31] Rick Chou, US and China Bully the Little Guy, *Taipei Times*, July 19, 1999, p.8.

[32] William J. Clinton, Transcript: President, First Lady on China in 21st Century, *USIS Washington File*, June 30, 1998. http://www.usia.gov/current/news/

[33] The *People's Daily* (oversea's edition), December 4, 1997. In this issue, the first point of Zemin's eight points (the so-called Jiang's eight points) was elaborated more than ever before. Beijing argued when the two sides across Taiwan Straits can coexist peacefully, this strategy would highlight a healthy direction for the development of cross-strait relations.

[34] The Journal *Cross-Strait Relations* (Beijing). December 10, 1997.

[35] *Wen Wei Po*, December 14, 1997.

[36] Rich Chou, US and China Bully the Little Guy, *Taipei Times*, July 19, 1999, p.8.

[37] Ling Cheng-yi, The Toughest Task for the New President, *Taipei Times*, July 1, 1999, p.4.

[38] *Taipei Times*, July 1, 1999, p.1.

Plate 4 Roof corner ➧

3

The Linkage of the ROC's Mainland China Policy and Foreign Policy

Marion C.Y. Wang

Associate Professor, Institute of Political Science, National Sun Yat-Sen University

The Republic of China (ROC) has been a sovereign state since its establishment in 1912. As a result of civil war and the founding of the People's Republic of China (PRC) on the Chinese mainland in 1949, Taiwan and the Chinese mainland have been separately governed. The ROC and the PRC conduct their own foreign affairs, exercise separate legal jurisdictions, implement their own administrative policies and enjoy economic autonomy. Neither is subordinate to the other.

The ROC's Mainland China Policy vs. The PRC's Taiwan Policy

The ROC's Mainland China Policy

The following elements have shaped the evolution of the ROC's mainland China policy since 1949: (1) the ROC's domestic politics; (2) international relations; and (3) the PRC's domestic politics, foreign policy, and Taiwan policy. The ROC's mainland China policy has ranged from "defend Taiwan, recover the mainland" via "one China, two entities, step-by-step equality, peaceful coexistence" to "special state-to-state relationship, one China after future democratic reunification."[1]

Since the Second World War, international relations have changed from Cold War to post-Cold War détente. In the post-Cold War era, three trends shape the new order of international community: namely, globalization, regionalism, and nationalism. In reaction to this new order, the late President Chiang Ching-Kuo in 1987 lifted the Emergency Decree. The ROC government terminated the "Period of National Mobilization for Suppression of the Communist Rebellion" and made constitutional amendments to more thoroughly implement constitutional democracy.[2] In the meantime the people on Taiwan were allowed to visit relatives in mainland China.

The National Unification Council was founded in October 1990 to manage the new situation of cross-strait relations and passed the "Guidelines for

National Unification" in February 1991, which maintain that China's unification should have as its goals the development and propagation of Chinese culture, safeguarding of human dignity, protection of basic human rights, and realization of the democratic rule of law. The non-governmental Straits Exchange Foundation (SEF), which is authorized by the ROC government, takes charge of the cross-strait exchange and dialogue.

The position of the ROC government is that, following an appropriate period of forthright exchange, cooperation, and consultation, the two sides of the Taiwan Straits can, with the presupposition of reason, peace, parity, and reciprocity, achieve a gradual reduction of hostilities. The ROC and the PRC can foster a consensus of democracy, freedom, and equal prosperity, abandon ideological entanglements, work together for the well-being of the entire Chinese population, and eventually achieve the goal of reunification in measured phases. The ROC government has not set a timetable for reunification, but "Taiwan Experience"[3] will play a key role in the process of national reunification.

In order to promote dialogue and exchange with the PRC, the ROC has taken steps to ease the situation, steps such as allowing people to visit relatives on the mainland of China, gradually reducing the restrictions on people-to-people exchanges and contacts, expanding indirect trade, and permitting indirect investment.[4] The ROC military buildup is oriented toward preventive defence.

An arms race, especially involving nuclear weapons, is incompatible with the ROC's national interest.[5] In 1991 the ROC government announced the adoption of a peaceful approach to achieving national reunification; even so, the PRC has not renounced the use of force toward the ROC.[6] Besides, in the process of the cross-strait exchange and dialogue there are some setbacks. These setbacks can be attributed to a cognitive problem concerning the definition of the "one China" principle.[7] In order to resolve grave differences of opinion over the meaning of "one China," representatives of the two sides reached consensus in November 1992 that each side would be allowed its own interpretation. As a result of this consensus, the Koo-Wang talks were held in Singapore in April 1993,[8] and four agreements were signed, including the Agreement on the Establishment of Systematic Liaison and Communication between the SEF and the Association for Relations Across the Taiwan Straits (ARATS). The institutionalized channels of consultation were therefore established.

The increase of people-to-people contacts and visits, and the rapid expansion of cross-strait trade and investment led to a number of important issues, such as:

- the safety of Taiwan residents in the mainland China area;
- protection of investments by Taiwan businessmen on the Chinese mainland;
- related customs and tariffs issues;
- handling of fishery disputes;

- repatriation of illegal entrants and joint efforts to combat crime; and
- mutual assistance concerning judicial matters.

All of these issues, as well as problems arising from contacts between the SEF and the ARATS, have been left unresolved.[9]

In order to break the impasse in cross-strait relations, in April 1995 the ROC's President Lee Teng-hui issued a six-point proposal to lay down a foundation for the development of cross-strait relations. The position of the ROC government is that the two sides can respect each other, find common ground while resolving their differences, increase cooperation, and enhance consensus on shared points of view. On the basis of the six-point proposal, the ROC has made a number of proposals, including the strengthening of cultural exchanges, developing complementary economic and trade contacts, assisting the Chinese mainland in improving agriculture, meetings in international settings between top leaders from both sides of the Straits, a "journey of peace" to the mainland by President Lee, exchanging experiences in reforming state-owned enterprises, and jointly inviting nations in Southeast Asia to discuss measures to reduce the impact of the Asian financial crisis. Both sides of the Straits should seek a formula and opportunities for cooperation with each other to jointly enhance the development and prosperity of the entire Asia-Pacific area. There has been no response from Beijing.

In recent years the PRC has tried to promote grassroots democracy, and it is in the ROC's interest that the PRC advance its political reforms and expand the scope and depth of its democracy.[10] After all, the pursuit of national democratization and modernization is the basis for the long-term, stable development of cross-strait relations. Democratization on the Chinese mainland and the positive development of cross-strait relations are, in turn, decisive factors in the process of China's reunification.[11]

In order to achieve the goal of national reunification, the ROC government passed more than 200 executive orders to open up private-sector exchanges between the two sides. During this time the people of the two sides have exchanged over 10 million visits and more than 100 million letters. Over 400 million phone calls have also been made.[12] The ROC government is eagerly hoping both sides can resume the cross-strait dialogue without conditions attached and as soon as possible. The ROC government reaffirms its opposition to the so-called "one country; two systems" formula. Taipei hopes to expand cross-strait relations and promote regional stability and development under the principles of peace, cooperation, and prosperity. In the Koo-Wang talk of October 18, 1998 the ROC had already extended an invitation to ARATS Chairman Wang Daohan to come to Taiwan in the Fall of 1999 for a visit. In order to manage an eventual high level political agenda (e.g., "one country; two systems") in coming Koo-Wang talks, President Lee Teng-hui asserted that the cross-strait relations are a "special state-to-state relationship."[13] The tension in the Taiwan Straits has therefore escalated.

The PRC's Taiwan Policy: One Country, Two Systems

The goal of the PRC's Taiwan policy is China's reunification, and the strategy to reach this goal has entered a new phase. After the founding of the PRC on the Chinese mainland in 1949, the strategy to complete China's reunification was the military liberation of Taiwan, until July 1955 when Zhou Enlai told the National Peoples Congress (NPC): "When conditions permit we hope to liberate Taiwan by peaceful means." On September 30, 1981, Ye Jianying, Chairman of the NPC Standing Committee, expounded on the "Guidelines on making Taiwan return to the motherland in a peaceful reunification" (commonly known as Ye's nine-point proposal). He affirmed that "after the country is reunified, Taiwan can enjoy a high degree of autonomy as a special administrative region" and proposed the talks could be held on an equal footing between the ruling parties on each side of the Straits, namely the Chinese Communist Party and the Kuomintang.

In January 1982, Deng Xiaoping pointed out that this in effect meant "one country, two systems," that is, on the premise of national reunification, the main body of the nation would continue with its socialist system while Taiwan could maintain capitalism. There was no timetable for Chinese reunification. On February 22, 1984 Deng elaborated: "After unification, China may adopt one China, two systems." The PRC has subsequently taken measures to manage bilateral exchanges and cooperation in areas, such as two-way travel, post and communications, as well as scientific, cultural, sports, academic, and journalistic activities. A non-governmental ARATS has been set up and authorized by the PRC government to liaise with the SEF and other relevant non-government bodies in Taiwan for the purpose of upholding the legitimate rights and interests of people on both sides and promoting cross-strait relations. On the military plane, initiatives have been taken to ease military confrontation across the Straits. The shelling of Jinmen (Quemoy) and other islands has been discontinued. Some forward defence positions and observation posts along the Fujian coast have been transformed into economic development zones or tourist attractions. Businesses from Taiwan are accorded preferential treatment and legal safeguards.[14]

On October 12, 1992, General Secretary of the Communist Party of China (CPC) Jiang Zemin pointed out: "We shall work steadfastly for the great cause, adhering to the principles of peaceful reunification and 'one country, two systems.'" The CPC is ready to establish contact with the Chinese Kuomintang at the earliest possible date to create conditions for talks on officially ending the state of hostility between the two sides of the Taiwan Straits and gradually realizing peaceful reunification. Representatives from other parties, mass organizations and all circles on both sides of the Taiwan Straits would be invited to join in such talks.[15] Beijing allowed low-level political and academic groups to make contact with the independence-oriented opposition party, the Democratic Progressive Party (DPP).

On January 30, 1995, PRC President Jiang Zemin delivered a speech aimed at Taiwan entitled "Issues on Developing Cross-Strait Relations and Advancing Peaceful Reunification at the Present Stage" (commonly known as Jiang's eight points). Jiang implied that "one China is the People's Republic of China," and solicited acceptance by Taiwan of the "one China" concept with no reservations.[16] Since June 1995 the PRC has reiterated "peaceful reunification" and "one country, two systems"[17] as the basis of its Taiwan policy. Premier Li Peng said on October 24, 1995 that "there is only one China in the world, Taiwan is a part of China, and the PRC is *de jure* the sole government representing China as a whole." The PRC government still rejects abandonment of the use of force toward the ROC, especially in cases of Taiwan's independence, involvement of foreign countries in Taiwan's independence, and Taiwan's production of nuclear weapons.[18]

From the PRC's point of view, the ROC talks about the necessity of a reunified China, but their actions are always a far cry from the PRC's principle of "one China." Taipei does attempt to prolong Taiwan's separation from the mainland, and refuses to hold talks on peaceful reunification. Taipei even sets up barriers to curb further development of cross-strait exchange and dialogue. The purpose of Lee Teng-hui's visit to the US was to establish "two Chinas."[19]

Subsequently, Beijing moved on two fronts on cross-strait relations: stalling the second round of the Koo-Wang talks as well as dialogue at all levels between the SEF and ARATS on the one hand, and conducting a smear-campaign and sabre-rattling via media attacks and missile tests on the other hand.[20] Koo Chen-Fu, Chairman of SEF, was invited to visit the PRC in October 1998. The tense atmosphere over the Taiwan Straits eased gradually. Meanwhile, President Lee proposed the concept of a "special state-to-state relationship." In reaction to President Lee's version, the PRC put pressure on the ROC to return to the "one China" principle with missile launching tests, military exercises, and diplomatic manoeuvres.[21]

The US Factor in Cross-Strait Relations

In early 1979, US President Jimmy Carter broke formal diplomatic ties with the ROC and then signed the Taiwan Relations Act (TRA) on April 10, 1979, creating domestic legal authority for the conduct of unofficial relations with Taiwan. The Act requires the US to make available to Taiwan defence articles and services in such quantity as may be necessary to enable Taiwan to maintain sufficient self-defence capacity. In a 1982 communiqué, the US stated that: it did not seek to carry out a long-term policy of arms sales to Taiwan, and that US arms sales would not exceed, either in qualitative or in quantitative terms, the level of those supplied in recent years. From the US point of view, maintaining diplomatic relations with the PRC has been recognized by six consecutive administrations to be in the long-term interest of the US. However, maintaining strong, unofficial relations with Taiwan is also in the interest of the US.[22]

President Clinton's attitude to cross-strait relations is that all issues should be resolved by peaceful means. He has also stated publicly that arms sales to Taiwan are not an obstacle to China's peaceful reunification. Meanwhile, in order to integrate the PRC into the international system, including the WTO, and transform the PRC's system peacefully, the Clinton administration carries out an engagement policy via strategic dialogue with Beijing, that is, mutual visits of the highest-level personnel.[23]

Geopolitics dictates that positive development of bilateral relations between the US and the PRC is in the interest of bilateral relations between the US and Taiwan, as well as bilateral relations between the PRC and Taiwan.[24] The position of the Clinton administration on cross-strait relations is to keep the "status quo"; "no independence, no force" and the "three no's" policy.[25] After the second Clinton-Jiang talks, the PRC and the US established a "constructive strategic partnership" in order to defuse cross-strait tension after the PRC's missile test toward the ROC in 1995-96. The Clinton administration urges not only the resumption of cross-strait talks, but also the signature of interim agreements.[26] Taipei regards urgings from the Clinton administration as pressure to undertake high-level political dialogue with Beijing. This pressure is one of the reasons that pushed President Lee to announce the concept of the "special state-to-state relationship" before the Koo-Wang talks in the fall of 1999.

Although there are differences between the PRC and the US regarding nuclear theft, the US-led NATO bombing of the PRC's embassy in Belgrade and WTO accession,[27] after the announcement of the ROC's President Lee's version of the "special state-to-state relationship," the US President responded with the three pillars of US China policy: namely, "one China, constructive dialogue, and peaceful resolution of differences between the PRC and Taiwan." The US government further asked the ROC to explain whether their mainland China policy has fundamentally changed. Meanwhile, US President Clinton used the hot line to contact PRC President Jiang Zemin about a peaceful resolution of the tense cross-strait relations.[28] The tense relations between the US and the PRC after the NATO bombing of the PRC's embassy in Belgrade are consequently back on track.

Clinton's statement on the "three no's" and "three pillars" policy has caused further isolation of the ROC's foreign relations from the international community. The ROC will urge the US to implement the TRA of 1979, the "six guarantees" of 1982, and the promises of the Taiwan Policy Review in 1994. The ROC hoped the US would dispatch cabinet-level officials to Taiwan as soon as possible to facilitate the balanced development of the "three sets of bilateral relations" among the ROC, the US, and the PRC.

THE DILEMMA OF THE ROC'S FOREIGN RELATIONS

PRC's Diplomatic Containment Against ROC 's International Activities

Recognizing the long-established influences of the US and Japan on the orientation of the ROC's international activities, the PRC has taken up contacts with the US and Japan to isolate the ROC in its international activities. In the Shanghai communiqué of 1972 the PRC and the US stated their respective positions. The PRC stated: "The Chinese (PRC) side reaffirms it position; Taiwan is a province of China which has long been returned to the motherland; the liberation of Taiwan is China's internal affairs."

Since establishing diplomatic ties with the PRC, the US has acknowledged that "all Chinese on either side of the Taiwan Straits maintain there is but one China and that Taiwan is a part of China." The US government does not challenge that position, and reaffirms its interest in a peaceful settlement of the Taiwan question by the Chinese themselves. Japan's position is outlined in the September 1972 Tanaka-Zhou communiqué, which stated: "The government of the PRC reaffirms that Taiwan is an inalienable part of the territory of the PRC. The government of Japan fully understands and respects this stand of the government of China and adheres to its stand of complying with Article 8 of the Potsdam Declaration."[29] After the establishment of diplomatic ties between the PRC and the US and Japan, the Third Plenary Session of the Eleventh Central Committee of the CPC decided to shift the focus of the work of the party and the state to the economic modernization programme, and to participate in the international community under the "five principles of peaceful coexistence;" namely, mutual respect for sovereignty and territorial integration, mutual nonaggression, non-interference in each other's international affairs, equal and mutual benefit, and peaceful coexistence.[30]

As a permanent member of the UN Security Council, the PRC's position concerning the ROC's foreign relations is that Taiwan, as a part of China, has no right to represent China in the international community; nor can it establish ties or enter into relations of an official nature with foreign countries. Beijing has not objected to non-governmental economic or cultural exchanges between Taiwan and foreign countries. All countries which have diplomatic relations with the PRC should abide by the "five principles of peaceful coexistence" and recognize that there is only "one China," that the PRC government is the sole legal government of China, and that Taiwan is part of China. They should also refrain from providing arms to Taiwan in any form or under any pretext.[31]

From the PRC's point of view, the ROC government has vigorously launched a campaign of "pragmatic diplomacy"[32] to cultivate official ties with countries having diplomatic relations with the PRC in an attempt to push "dual recognition" and achieve the objective of creating a situation of "two Chinas" or "One China, one Taiwan." Besides, Taipei's lobbying for a formula of "one country, two seats" in international organizations whose membership is confined to sovereign states is also a manoeuvre to create "two Chinas."

The PRC government allows the ROC's participation in the activities of international organizations only on the premise of adhering to the principle of "one China" and in light of the statutes of the international organizations concerned, as well as the specific circumstances. As regards participation in nongovernmental international organizations, the relevant bodies of the PRC may reach an agreement or understanding with the parties concerned, so that the PRC's national organizations would use the designation of China, while the ROC's organizations may participate under the designation of Taipei, China or Taiwan, China.

As to regional economic organizations, such as the Asian Development Bank (ADB) and the Asia-Pacific Economic Cooperation Conference (APEC), the ROC's participation is subject to the terms of agreement or understanding reached between the PRC and the parties concerned, which explicitly state that the PRC is a full member as a sovereign state, whereas Taiwan may participate in the activities of those organizations only as a region of China under the designation of Taipei, China (in ADB) or Chinese Taipei (in APEC). From the PRC's standpoint, this is only an ad hoc arrangement and cannot constitute a "model" applicable to other inter-governmental organizations or international gatherings.[33]

After President Lee's visit to the US as a Cornell alumnus, and Vice-President Lien Chan's visit to Europe, the PRC has begun to restrict the ROC's diplomatic activities closely. The PRC has used the pretext that Taiwan's "pragmatic diplomacy" is in essence creating "two Chinas" or "one China, one Taiwan,"[34] and has threatened to roll back its relations with Washington in an attempt to compel the US to pull back from its current relation with the ROC.

Further, the PRC has claimed that it would reduce the number of the ROC's allies to zero. It has adopted the so-called "three zeros" policy; namely, zero allies for Taiwan, zero international space for Taiwan, and zero bargaining chips for Taiwan to negotiate with the PRC. On the one hand, it focuses on the US, Japan, and the advanced countries in Europe, trying to force them to express their support for its "one China" policy, not to support Taiwan's bid to participate in the United Nations (UN), and not to sell advanced weapons to Taiwan. After the US-led NATO bombing of the PRC's embassy in Belgrade on May 7, 1999 one of the PRC's premises of developing relations with the US is to exclude the ROC from participation in the Theater Missile Defense (TMD).[35] On the other hand, it lists Africa, Central and South America, and the South Pacific as the major battlefields of struggles with the ROC. It utilizes huge amounts of financial aid and its capacity as a permanent member of the UN Security Council to sabotage the ROC 's relations with allies.[36]

PRC Vice-Premier Qian Qichen visited five countries in Central and South America in July 1997. During his trip, Qian stated publicly that the PRC hoped to develop relationships with those countries having diplomatic ties with the ROC. In addition, Qian tried to contain the ROC's efforts in this area, taking advantage of US President Clinton's remark concerning the "three no's" policy. Furthermore, the PRC invited members of both ruling and opposition parties in the Caribbean countries having diplomatic ties with the ROC to the conference of friendship with the PRC held in Jamaica at the end of July 1997. After the announcement of President Lee's version of the "special state-to-state relationship," Qian stated that the PRC's containment against the ROC's international activities will be absolute; even so the US, Japan, the EU, and ASEAN countries repeated their "one China" position.[37]

The Limitation of the ROC's International Activities

The ROC's foreign policy changed with its aforementioned mainland China policy and international relations. In the Cold War era, the ROC government insisted that the ROC was *de jure* the sole representative of China in the international community and cooperated with the US and Japan in international affairs. In the post-Cold War era the ROC recognized that its *de facto* control extended only over Taiwan, Penghu, and Matsu.[38] In order to participate in the international community actively, the "pragmatic diplomacy" approach is put into practice by the ROC government, which is definitely not aimed at "Taiwan Independence."

The PRC's obstructionism is the only reason that other states have not recognized the ROC as a sovereign state. From the ROC's viewpoint, only when its international status is ensured can it begin talks with the PRC on an equal footing over the issue of national reunification. To become a member of the UN is one of the ROC's major foreign policy objectives. From the ROC's viewpoint, in the post-Cold War era, resolution 2758 (XXVI) of the UN General Assembly does not constitute a comprehensive, reasonable, and just solution to the question of representation of the Chinese people in the UN.[39]

The following cases indicate that, although the ROC has contributed to the international community, it is still not allowed to take part in the UN program:

(1) The ROC set up the International Economic Cooperation Development Fund (IECD) in 1988, which, up to March of 1995, has committed 23 loans valued at US$310 million to 16 countries. The ROC is a major contributor to the ADB. In 1991, the ROC and the European Bank for Reconstruction and Development (EBRD) signed an agreement to establish the Taipei China-European Bank Cooperation Fund in the amount of US$10 million to finance technical assistance, training, and consultation services for reconstruction works in Central and Eastern Europe.

    The ROC is also involved actively with the Inter-American Development Bank (IDB), and it entered into co-financing and re-lending cooperation with the Central American Bank for Economic Integration (CABEI), of which it is a non-regional member. The ROC has sent 45 technical cooperation missions to more than 30 countries around the world, providing medical, agricultural, fishery, and handicraft assistance. Since 1962, the ROC has also offered vocational training courses in agriculture, land reform, industrial technique, trade, small and medium enterprises development, taxation, customs, and scientific technology to approximately 7,500 trainees from 80 countries. But the ROC is still unable to join the development programs sponsored by the United Nations Development Program (UNDP) and it is excluded from the activities of the International Monetary Fund (IMF) and the World Bank Group.[40]

(2) Since 1981 the ROC has participated actively in international disaster relief and humanitarian aid operations. In 1990, the ROC government institutionalized its support for such operations by establishing the "international disaster relief aid fund." The government's annual budgetary process provides the allocation for the fund. From 1990 to March 1995 the ROC has directly or indirectly provided disaster relief and humanitarian aid of US$124 million to more than 60 countries. In 1999 the ROC government provided humanitarian aid of US$ 30 million to Kosovo refugees. In addition, private charity organizations in the ROC have also supported international medical assistance programs and anti-poverty campaigns with financial and human resources. But the ROC is still unable to participate in the operations of UN agencies, such as the UN High Commission for Refugees (UNHCR) and the UN Children's Fund.

(3) The ROC was prevented from becoming a contracting party in international conventions on ecological and environmental protection, such as the United Nations Framework Convention on Climate Change and the Montreal Protocol on Substances that deplete the ozone layer. The ROC has taken measures in ecological and environmental protection in keeping with the aforementioned international conventions, but was forced to live under the shadow of trade sanctions imposed by the protocol.[41]

The ROC has striven for admission to the UN on the basis of the following three premises:

(a) The ROC continues to pursue the eventual reunification of China;

(b) Taipei's application is not meant to challenge the current seat of the PRC in the UN; and

(c) The campaign recognizes the fact that the two sides of divided China exercise jurisdiction over separate territories, and seeks for the 22 million people under the jurisdiction of the ROC a proper and effective protection and representation of their fundamental rights in the UN.

The UN, through the PRC's veto, rejected the ROC's applications for UN admission. In addition, although the ROC participates in APEC and has made efforts to maintain ties with allies in the Asia-Pacific area via agricultural technological cooperation, economic aid, and invitations for visits, as well as the plan for an Asia-Pacific Regional Operation Center,[42] the scope and depth of the ROC's activities in the Asia-Pacific area still depend on the PRC's attitude.

CONCLUSION

The national power of a country is the basis on which to pursue its national interest in the international community. In the post-Cold War era the ROC government has taken steps, such as upgrading technical know-how, further democratization in domestic politics, increases in international aid, and the buildup of preventive defence, to advance its national power and achieve its national interests in cross-strait relations and in the international community. The key obstacle to the ROC's admission to international organizations (e.g., UN, WTO), and to the establishment of diplomatic ties with other countries is the PRC's "one China" principle. The ROC's strategy to break through the diplomatic impasse is to link its sovereignty-oriented foreign policy to the national reunification policy of mainland China.

ENDNOTES

1 *China Times,* July 21, 1999, p2.

2 The Emergency Decree activated martial law as well as a ban on the organization of new political parties, and threatened the registration of newspapers, in: Lee Teng-hui, *The Taiwan Experience and China's Future*—Address to the American Chamber of Commerce in Taipei and the American University Club, July 6, 1991, p. 2 (http://www.cop.gov.tw/presid/statement/abroad/e4-1.htm).

3 "Taiwan Experience" is based on the guidelines of the Three Principles of People, and entails being united and unwavering in the tasks of combining tradition with modernity, balancing idealism and pragmatism, working to achieve national progress, and striving to ensure the dignity of each individual." Our basis was the true spirit of the Three Principles of the People: ethics, democracy, and science. We upheld the basic tenets of the Three Principles of the People: freedom, democracy, and equitable distribution of wealth," in: Ibid.

4 Relations Across Taiwan Straits: Evolution and Stumbling Blocks, in: *The Taiwan Question and Reunification of China* , Taiwan Affairs Office & Information Office, State Council, The People's Republic of China, August 1993, Beijing (http://www.China.org.cn/white papers).

5 Since 1989, its official defense budget—not to mention actual expenditure—has grown at a double digit rate, far outstripping the inflation rate. Its post-strait-crisis military exercises, as frequent as previously, have included sea crossing, airborne activity, urban warfare, and subtropical jungle operations. These are expressly targeted at the ROC. Beijing's vigorous arms acquisitions, both through indigenous and foreign sources, have progressed rapidly, in Taipei on TMD Will the Republic of China Participate in the Theater Missile Defense? (http://www.mac.gov.tw/).

6 Mainland policy and the current state of cross-strait relations: The Republic of China's stance and approach, June 1998 (http://ww.gio.gov.tw/info/news/).

7 Koo, Chen-fu: Remarks at the Meeting with Mr. Jiang Zemin, October 18, 1998, p. 1 (http://www.mofa.gov.tw/emofa/koo1018be.htm).

8 *Mainland policy and the current state of cross-strait relations,* loc. cit.

[9] Koo: Key points from remarks made at a meeting with ARATS Chairman Wang Daohan, Oct. 12, 1998, p. 1 (http://www.mofa.gov.tw/emofa/koo 1012ae. htm).

[10] Koo: Remarks at the meeting with Mr. Jiang Zemin, loc. cit.

[11] Koo: Key points from remarks made at a meeting with ARATS Chairman Wang Daohan, loc. cit.

[12] *Mainland policy and the current state of cross-strait relations*, loc. cit.

[13] *China Times*, July 10, 1999, p. 1.

[14] Relations across Taiwan Straits: Evolution and stumbling blocks, in: *The Taiwan Question and Reunification of China*, Taiwan Affairs Office & Information Office State, Council of The People's Republic of China, August 1993, Beijing, p 1 (http://www.china.org.cn/ white papers/reunification of China E.2 html.).

[15] *The Chinese Government's basic position regarding settlement of the Taiwan question*, ibid.

[16] A Preliminary Analysis of Mainland China "One China Strategy," p. 1 (http:// www.mac.gov.tw/).

[17] Basic contents of "peaceful reunification" and "One country, two systems" are as follows: 1) Only one China: Taiwan is an inalienable part of China. Self -determination for Taiwan Independence is out of the question; 2) Coexistence of the two systems—on the premise of one China, socialism on the mainland and capitalism on Taiwan can coexist and develop side by side without one swallowing up the other; 3) A high degree of autonomy—after reunification, Taiwan will become a special administrative region, in: *The Taiwan Question and Reunification of China*, loc. cit.

[18] *China Times*, July 20, 1999, p. 14.

[19] Relations Across Taiwan Straits: Evolution and Stumbling Blocks, ibid.

[20] A Preliminary Analysis of Mainland China's "One China strategy," loc. cit.

[21] *China Times*, July 15, 1999, p. 1.

[22] USIA, US Foreign Policy Agenda, January, 1998—fact sheet: US Asia-Pacific Security Alliances (http://www.usia.gov/journals,itpa/0198/ijpe/pji/8fact. htm).

[23] Washington-Taipei Relations and ROC's Pragmatic Diplomacy: Question and Answers, May 1997, p2 (http://www.mofa.gov.tw/).

[24] Oksenberg, Michel, *China Times*, September 01, 1999, p. 14.

[25] The US "three no's" policy means that the US does not support "Taiwan independence," "two Chinas, one China, one Taiwan," or Taiwan's participation in international organizations whose membership is confined to sovereign states," in: *China Times*, July16, 1999, p. 2.

[26] *China Times*, July 1, 1999, p. 1.

[27] Sino-US Relations after Zhu's Visit (http://www.china.org.cn/bjreview/99 may/bjr99-20e-04.html).

[28] *China Times*, July 23, 1999, p. 2.

[29] The Cairo Declaration issued by China, the United States and Great Britain in December, 1943 stated: "It is the purpose of the three great Alliances that Japan shall be stripped of all the islands in the Pacific which it has seized or occupied since the beginning of the First World War, and that all the territories Japan has stolen from the Chinese, such as Manchuria, Formosa (Taiwan) and the Pescadores (Penghu), shall be restored to China." The Potsdam Proclamation signed by China, the United States and Great Britain on July

26, 1945 (subsequently adhered to by the Soviet Union) reiterated: "The terms of the Cairo Declaration shall be carried out," in: *Washington-Taipei Relations and ROC's Pragmatic Diplomacy: Question and Answers*, op. cit., pp. 1-2.

30 Arms sales to Taiwan by countries having diplomatic relations with China, in: *The Taiwan Question and Reunification of China*, op. cit., p. 1.

31 Several questions involving Taiwan in international relations; arms sales to Taiwan by countries having diplomatic relations with China, in: *The Taiwan Question and Reunification of China*, op. cit., p. 1 and p. 2.

32 "Pragmatic diplomacy" is the term used to describe Taiwan's wide-ranging effort to take its rightful place in world affairs, in: *It's time to be pragmatic* (http://www.gio.gov.tw/info/UN50/pragmatic html).

33 Several questions involving Taiwan in international relations, in: *The Taiwan Question and Reunification of China*, loc. cit.

34 A Preliminary Analysis of Mainland China's "One China Strategy," loc. cit.

35 *China Times*, May 13, 1999, p. 14.

36 Hu, Jason C. *The current state of ROC diplomacy*. An Abridgment of the Report to the Foreign and Overseas Chinese Affairs Committee Legislative Yuan, September 21, 1998, p. 1.

37 *China Times*, July 16, 1999, p. 2; July 14, 1999, p. 3; July 21, 1999, p. 14; July 25, 1999, p. 13.

38 As a sovereign state since 1912, the ROC meets all criteria of the Montevideo Convention on rights and duties of states (1933). Article 1 of the Montevideo Convention on Rights and Duties of States defines a state as possessing a permanent population, a defined territory, a government, and the capacity to enter into relations with other states. Article 3 of the same Convention says that "the political existence of the state is independent of recognition by other states, in: *Washington-Taipei Relations and ROC's Pragmatic Diplomacy: Questions and Answers*, loc. cit.

39 The Fundamental Rights of the People and Government of the ROC on Taiwan to participate in the United Nations and international activities (http://www. gio.gov.tw./info/UN50/tie/html), p. 3.

40 Ibid. p. 1.

41 Ibid.

42 The Plan is divided into two parts; a macroeconomic adjustment program and a specialized operations centre development. The macroeconomic adjustment program is designed to improve Taiwan's overall economic infrastructure by allowing the "four I's" (investment, industry, individuals, and information) to flow freely in and out of Taiwan, and the operations centre is a centre for six economic activities in the Asia-Pacific: manufacturing, sea transportation, air transportation, financial services, telecommunications, and media production, in: *Standing on Taiwan, Embracing the Asia-Pacific Region* (http:www.gio.gov.tw/info/un50/un03.html); Lee, Teng-hui: Remarks at the Dinner in Honor of the Diplomatic Corps in the Republic of China, February 6, 1998 (http:www.oop.gov.tw/).

Plate 5 Entrance to east-west, cross-island highway ➧

東西橫貫公路
開放 OPEN

Arms Control and Disarmament in East Asia—The Korean Peninsula and the Taiwan Straits

4

Robert E. Bedeski
Professor, Department of Political Science, University of Victoria

Strategic Overview—The Need for Tension Reduction in the Asian Region

Beginning in the mid-19th century (Chinese historians would start with the Opium Wars), a historical source of tension and conflict has been the strategic imbalance between China and Japan. Several false starts in modernization by China, the collapse of the imperial Confucian system, and subsequent fragmentation under warlords and a Potemkin Republic, consigned China as the sick man of Asia, unprotected against Japanese hegemony. Japan had seen the writing on the wall—that the old order of East Asia with China at the centre was collapsing, and a new order of Western nation-states, based on law and limited democracy, and propelled outward by science and Christianity, was to dominate the globe. As Japan modernized and militarized, learning that empire was the right of modern nation-states if only they seize the opportunity, Korea, and then China, became lusted territory. The security deficit of China against Japan increased, and by the 1930s, Chiang Kai-shek was forced to trade Chinese territory for time—until the US entered the war against Japan, and defeated her, not from Chinese bases but from the Pacific island-hopping campaign which provided the unsinkable aircraft carriers to take the war to the Japanese heartland.

Communist revolution followed World War II, but there was no relief from security concerns. Fears of Japanese remilitarization partly prompted Mao to travel to Moscow, and sign the Sino-Soviet alliance in January 1950. Before the end of the year, Communist China found itself at war in Korea against the US and UN forces, after Kim Il Sung's war of liberation into the south was thrown back. With the signing of the Military Armistice Agreement in 1953, the stalemate on the Korean Peninsula has persisted for nearly half a century, with no sign of resolution. The collapse of the Soviet Union brought little relief, since an even more isolated Democratic People's Republic of Korea (DPRK) has pursued development programs of weapons of mass destruction (WMD), and threatened the stability of the region with missile development.

During the Cold War, the alignments of the Korean War continued, with the People's Republic of China (PRC), United Soviet Socialist Republic (USSR), and North Korea and their allies aligned against the US and its allies. More than four decades of Cold War saw nuclear arms races, and entrants of new nuclear players—including India and the PRC. The wars in Indo-China smouldered and blazed from the mid-1950s until 1975. Nixon-Kissinger diplomacy took advantage of the Sino-Soviet rift, and set in motion the recognition of the Beijing regime in 1979.

From the reluctant power that had to be shocked by Pearl Harbor to go to war in the Pacific, the US became, by the end of the war, the major superpower and the Pacific an American lake, with no challengers to its dominance. With power came responsibility, and to carry out its self-imposed mission, the US formed a number of bilateral alliances, and sought to balance the influence and power of China and the Soviet Union. With the collapse of the Cold War system, the US has continued its balancing role, seeing China as the most likely hegemon in the region to challenge US dominance. After ending the base leases from the Philippines, the main area of concern has been the north Pacific, and particularly Taiwan, South Korea, and Japan.

Since 1993 North Korea has been a central concern, prompting the revival of anti-missile defence, involving Japan and possibly Taiwan. The collapse of the Soviet Union in 1991 left North Korea isolated, and shifted the centre of Communist opposition to the US in the PRC. Both countries have opened up to each other, with trade, cultural exchanges, and investment, but issues such as human rights, Taiwan, and military deployments persist as irritants and sources of mutual suspicion.

Tensions in Post-Cold War Asia

The end of the Cold War has reduced general global tensions, but this was offset in Asia by continued disputes and conflicts between India and Pakistan, the PRC and ROC, and DPRK and Republic of Korea (ROK). Increased arms spending and military modernization have accompanied economic growth in Northeast and Southeast Asia.

> *In the case of arms imports to the region, Asia's share of world expenditure on arms transfers rose from 15.5% in 1982 to 34% in 1991. In 1991, three countries in the Asia Pacific region—South Korea, China, and Thailand—ranked in the top 10 arms importers in terms of contracts concluded; two others— Taiwan and Burma —ranked in the top 10 in terms of the value of arms actually delivered.*[1]

Economic crisis in Asia weakened US allies—especially South Korea and Japan—but the region appears to be well on the road to recovery. The US is the only global superpower today, and is at the most prosperous point in history. Although critics claimed that Congressional defeat of CTBT was a symptom of isolationism, evidence for a general retreat from global activism is not there. The US has engaged in measured responses to a number of post-Communist crises, including several in Yugoslavia, East Timor, and elsewhere. The US has taken the initiative on the question of North Korea, and has reiterated determination to maintain peace in the Taiwan Straits.

The reality of peace and stability in East Asia is that military power remains the key factor, and the US will be the primary guarantor in the region. This, of course, does not absolve smaller countries from responsibility for their own defence, and resolution of outstanding issues. ASEAN and the Asian Regional Forum (ARF) provide frameworks for regional cooperation, but thus far there is no parallel organization in Northeast Asia. The major problem for Northeast Asia today is to prevent confrontation and conflict, or to moderate confrontation when it does occur. Diplomacy and conflict mediation provide important potential tools, but the problem of weapons proliferation is more immediate and must be dealt with as early as possible.

Arms Control and Disarmament—Important Differences

Arms control may provide one route for dealing with the weapons question and providing the foundation for resolution of other issues by bringing together antagonistic parties to discuss technical military issues, establishing a pattern of dialogue, experimenting with confidence building measures (CBM), and perhaps moving to other issues. The process takes years, even decades, before results are visible, as can be seen in the record of US-Soviet negotiations.

There are two distinct approaches to the problem of weapons, although these are often lumped together as a single process. The major assumption behind arms control and disarmament efforts is that weapons contribute to the outbreak of war. This contrasts with the view that peace and stability can only be attained through balances of power or through preparation for war.[2] Those who advocate arms control and disarmament argue that eliminating weapons that pose threats can reduce the frequency of war. It is necessary to prevent arms races that increase tensions and hostility and absorb financial resources, and to promote mutual trust and confidence. Finally, the destructiveness of war must be limited when it occurs.

Disarmament seeks to drastically reduce, or completely eliminate, all weapons as a step in eliminating war itself. *Arms control*, on the other hand, aims to regulate the growth of weapons and reduce arms levels. The difference is between what is ideally desirable and what is realistically possible. *Arms control* aims not to eliminate weapons or war, but to reduce the risk of war through efforts to stabilize the status quo, build confidence between states and groups, encourage the peaceful resolution of disputes, and discourage the use of force.

Arms control is fundamentally a conservative enterprise. *Disarmament* seeks to overturn the status quo; *arms control* works to perpetuate it.[3]

Those who argue for disarmament may claim the moral high ground, and we can see this logic at work within Canada and the US. Gun registration and control is a step toward domestic disarmament, but does it reduce violence? Weapons are a deterrent in an imperfect world. A recent book by John R. Lott Jr.[4] argues that citizen legal possession of guns plays a significant role in preventing crime. He says that the chances of serious injury from an attack are 2.5 times greater for women offering no resistance than for those resisting with a gun.[5] He also argues that homicide numbers are lower in localities where concealed guns are legal—criminals are less likely to attack homes or citizens if there is a probability of armed response.

The point is that eliminating weapons may be a high moral good, but weapons nevertheless can play a useful function in the flawed world we live in. Universal disarmament will in practice disarm only those who volunteer to participate—the congenital law-abiders—and make them vulnerable to those who fail to disarm. As to the argument that disarmament will make a nation peaceable, it should be recalled that by 1900 private firearms and other weapons were practically nonexistent in Japan due to confiscation. Few would argue that Japan became a peace-loving nation as a result.

The assumption of arms control is that disarmament is improbable, but that we can make the world safer through compromises, tension-reduction, and greater transparency. As long as there are conflicts and tensions, nations will resort to arms to defend themselves and their interests. Disarmament within a nation requires intrusive and authoritarian measures, and most democratic nations resort to half-measures that fail to achieve their objective, while interfering with what many citizens feel are their inalienable rights of self-defence. Externally, governments are reluctant to give up the weapons they feel necessary to defend their sovereignty. Another problem that has emerged in the wake of the Cold War is the proliferation of small weapons, as well as the deployed land mines left from past hostilities. The usual victims and casualties are the innocent civilians who live and travel on the former war frontiers and battlegrounds. Canada and other nations have taken the lead in this sector of human security, and most governments grant that land mines are not vital to national security, and therefore are amenable to disarmament measures.

The following sections examine some of the progress and proposals regarding arms control, arms reduction, and disarmament in three Asian hot spots—the Korean Peninsula, Taiwan Straits, and South Asia.

The Korean Peninsula

From 1991 through mid-1994, the US suspected that North Korean nuclear reactors enabled diversion of fissile material to weapons development, and so they increased the pressure to halt alleged nuclear weapons development.

The US came close to going to war again with the DPRK. Subsequent to Jimmy Carter's visit to Kim Il Sung in the summer of 1994, an agreement was signed in October 1994 setting up the Korean Peninsula Development Organization (KEDO), which provided financing and technology for North Korean light water reactors.

In 1998, North Korea tested a long-range missile, launching the Taepodong in August 1998 that flew over Japan and crashed in the North Pacific. Signs of construction for new weapons facilities were detected, and the US demanded inspections—for which the North Koreans countered with demands of $300 million for a peek.

The situation remains critical because of North Korean belligerency and desperation, driven in part by famine and bankruptcy, as well as by isolation. Frequent predictions of imminent collapse have been exaggerated, and Pyongyang has developed brinkmanship to a fine art. The dilemma of North Korea as a rogue state is the following:

- Its leadership often seems impervious to normal diplomatic carrots and sticks, although a case can be made that Pyongyang has in fact responded reasonably.
- It is a potential nuclear power developing long-range missile capability.
- It exercises totalitarian control over its population.
- It has carried out threats in the past, and might not back down if forced into a corner.
- It threatens the two major allies of the US in Northeast Asia, and thus challenges US guarantees of security.
- It has a sophisticated understanding of US politics, probably recognizes the limits of American willingness to engage in confrontation and, similar to Iraq, knows the weak points in the alliance confronting it.

In this setting, the normal tools of arms control and verification, and confidence and security building measures have little chance of seeing the light of day. KEDO may be the best compromise, and does offer a continuing, though expensive, link between the West and North Korea—as long as it delivers the prospect of cheap dependable nuclear energy to North Korea, and the end of nuclear weapons development for the KEDO consortium.

In short, the prospects for arms control or disarmament are not bright. The Korean war remains stalemated under the Military Armistice Agreement, and both Koreas mistrust each other to an extreme degree, while any progress has been made only in the face of major concessions to the DPRK.

Taiwan Straits

When China conducted missile tests near Taiwan in March 1996, it reminded the world of tensions there. From the ROC's perspective, nuclear weapons must be an attractive deterrent to further coercion from the PRC. When India and Pakistan conducted nuclear tests in 1998, voices in Taiwan urged the government to undertake its own nuclear weapons program, arguing that nuclear weapons were the best guarantees of survival. So far the ROC has not gone in this direction, because the costs would far exceed benefits, and nuclear weapons would probably complicate the relations the island enjoys with the US, Japan, South Korea, and Southeast Asia. Reliable friends are a better guarantee of security than nuclear weapons.

Beijing refuses to renounce the use of force, and is planning more missile tests, so there is pressure on Taipei to respond in some fashion. The US plans to include Taiwan in the theatre missile defence system (TMD) with Japan and South Korea. This will enable Taiwan to purchase, research, and expand high-tech weaponry—a development that China sees as destabilizing to the region.

One suggestion is that there should be establishment of a nuclear-free zone (NFZ) in the Taiwan Straits.[6] The concept of NFZ began in 1959, when 42 nations signed the Antarctic treaty, which was followed by the Treaty for the Prohibition of Nuclear Weapons in Latin America and the Caribbean in 1967, and the South Pacific Nuclear-Free Zone Treaty in 1985. Similar treaties covering Southeast Asia and Africa were signed in 1995 and 1996. The Northern hemisphere has been much slower to eliminate nuclear weapons, and Professor Hsieh Shu-yuan suggests that a modest start could be made by initiating a treaty covering the Taiwan Strait, with the US, China, Russia, South Korea, Japan, the ROC, and countries of Southeast Asia. Its primary purpose would be to contain the spread of nuclear weapons, and to compel those states possessing them to eventually abandon these arsenals. The key elements of a treaty would be to ban possession of nuclear weapons within the zone by any state, and the use or threatened use of nuclear weapons against targets in the zone. The hope is that this would lead to more transparency and confidence among the regional powers.

The solution is reasonable, but is unlikely for several reasons:

- The PRC will not sign any treaty with Taiwan, which it regards as a straying province.
- The PRC is not likely to renounce the use of force of any kind against Taiwan.
- The PRC sees any treaty with Taiwan as a signatory to be a step toward international re-recognition. In 1999, China exercised its rarely invoked veto in the UN Security Council to end UN peacekeeping in Macedonia because that country had normalized relations with Taiwan.

As in the case of the Korean Peninsula, the outlook for arms control and disarmament is not optimistic. In both instances, divided countries and incomplete sovereignty combined with the fact of unfinished civil wars render peaceful solutions most difficult. In the final case of India and Pakistan, the intensity of crisis is also related to the fact of incomplete territory—incorporation of Kashmir, claimed by both countries.

South Asia—The Nuclear Threat

Nuclear tests by India and Pakistan in May 1998 raised a number of challenges to nuclear security in general, and to the existing nonproliferation regime in particular. Their actions are seen as threats to the Nuclear Proliferation Treaty (NPT) (signed in 1968, and extended indefinitely in 1995). Many experts oppose recognizing India and Pakistan as legitimate nuclear states, and treat them differently from existing nuclear states.[7] India holds a key to the future of nuclear weapons, and argues that the NPT without provisions for nuclear disarmament only consolidates the status of nuclear states and denies equal sovereign rights to other states.

From this perspective, India's participation in what it sees as the unfair NPT regime would endanger national security by downgrading India's international status, and by threatening its strategy toward nuclear China. India sees China as its major threat, since its defeat in the 1962 border war. India therefore does not recognize the NPT as a just or an effective regime. Ironically, India follows the same logic as China in 1964, when it tested its first nuclear bomb—to develop nuclear weapons for its own defence and to proclaim commitment to global disarmament as the only acceptable method of achieving world peace.

Signatories to the NPT have chosen international stability over international equality. India's argument could also be used by Japan in the future to justify nuclear weapons—it has the capability, and if abandoned by the US would have ample reason to go nuclear, with North Korea and nuclear China as neighbours.

Indo-Pakistan conflict over Kashmir flared up again in May and June, 1999, reminding the world that the threat of war lurks just below the surface. Since both had conducted nuclear testing in the previous year, the situation becomes even more ominous. India chose to test in 1998 in part because the Bharatiya Janata Party (BJP) brought a radically new vision to India's security policy when it came to power. This vision departed from cautious policies of the past and placed greater emphasis on the long-term Chinese threat. This testing stimulated Pakistan to conduct tests of its own, in part for domestic reasons. Both countries have also tested ballistic missiles. Although the tests were not a revelation of new capabilities, they were a challenge to the international NPT regime. Some Indian experts indicated that the tests were a challenge to US hypocrisy over its possession of weapons, while seeking to deny other states that possession.

But a more fundamental shift may be taking place in India—a shift from the past strategic culture, which was cautious and failed to think of the country as a strategic entity, to a more decisive direction. Foreign Minister Jaswant Singh has clarified BJP aspirations:

> *...unless India gives some definition to its vital national interests, it will fail to even conceptualize its strategic frontiers. Thereafter, a*

> *violation of any of those interests will go entirely unchecked. In consequence, India's difficulties will be enhanced, future correctives will be made more difficultly, and the country's national security will be adversely affected... this critical deficiency lies at the heart of its present immobility, both of thought and of action.*[8]

In sum, India's new strategic logic is not only a disturbing symptom of perennial nuclear threats, but reinforces a renewed and convincing argument (to potentially nuclear states) for nuclear development in the Third World. Rogue states of the world—including China in the 1960s, North Korea today, Iraq, and others—have long ignored world opinion in pursuit of the ultimate WMD. India's arguments, however, carry much more weight, and may be more destructive to the NPT regime than the 1993 challenge from North Korea. India's testing has made Asia a more dangerous region.

Conclusions

The end of the Cold War has not eliminated the threat from nuclear weapons. The Framework Agreement of 1994, which set up KEDO, and the South Asian nuclear explosions in 1998 demonstrated that nuclear weapons are powerful bargaining chips as well as symbols of national power. While treaties such as the NPT and CTBT have a major role to play in arms control, and much progress has been made in Europe and the Southern hemisphere, the spectre of nuclear weapons remains a serious problem in the Middle East, South Asia, and Northeast Asia. Nuclear disarmament will remain a distant hope, and unilateral disarmament under such conditions would be folly as long as the technology to create these weapons remains accessible.

This leaves arms control as the best available means of dealing with these problems. China, India, and Pakistan must be engaged to deter them from further nuclear development, while security guarantees for Japan, South Korea, and Taiwan must be provided to discourage them from pursuing nuclear weapons development. China must take greater responsibility for discouraging North Korea from nuclear and missile programs, if Beijing fears the consequences of TMD. The US is trying harder to pursue a cooperative policy with North Korea. Without some creative steps, the slippery slope of nuclear proliferation will appear on the Asia Pacific horizon and we will all be the losers.

ENDNOTES

1 Ball, D. (1993/94). Arms and affluence: Military acquisitions in the Asia-Pacific region. *International Security*, 18(3), 78-112.

2 Sens, A., and Stoett, P. (1998). *Global politics: Origins, currents, directions.* Toronto: ITP Nelson, p. 254.

3 Ibid., p. 255.

4 Lott, J.R. Jr. (1998). *More guns, less crime: Understanding crime and gun control laws.* Chicago: University of Chicago Press.

5 Lott, J.R. Jr. (1998). Gun shy, *National Review*, December 21, p. 47.

6 Hsieh Shu-yuan (1999). Making a nuclear-free zone for the Taiwan Strait region. *The Free China Journal*, January 29, p. 6.

7 *Look Japan*, May 1999, p. 22.

8 Joeck, N. (1999). Nuclear developments in India and Pakistan, *Access Asia Review*, July, 2(2), 12-13.

Plate 6 A section of the east-west cross-island highway ➧

5

Globalization vs. Marginalization: The Experience of Taiwanese Urban Aborigines

Jou-juo Chu

Executive Director, Institute of Public Opinion Research, and Associate Professor, Sun Yat-Sen Institute of Interdisciplinary Studies, National Sun Yat-Sen University

Introduction

Today, Taiwan has nine major tribes of aboriginal people. They are the Atayal, Saisiyat, Bunun, Tsou, Paiwan, Rukai, Puyuma, Ami, and Yami. The total population of Taiwan's aborigines currently amounts to 370,000, or about 1.73% of the total population. The Ami and Puyuma tribes have the largest plains-dwelling population. Almost 94% of Puyuma and 71% of Ami tribal people are concentrated in Hualian and Taitung, two agricultural counties in East Taiwan. In addition to these, the Saisiyat tribe has 48% of its population, and the Rukai tribe has a quarter of its population, scattered across the aborigine towns in the flat plain area. The aboriginal tribes that still maintain a large mountain-dwelling population are the Yami and Tsou. The former has almost 95% of its population and the latter 91% concentrated in mountainous areas. The Atayal and Paiwan claim to have the third and fourth largest population of mountain settlers. In comparative terms, the aboriginal tribes with larger plain settlements have a higher ratio of urban migration than their mountain counterparts (Chen and Sun, 1994, 50-51).

The concept of urban aborigines here refers to those aborigines who have migrated to the cities and adjusted themselves to an urban lifestyle. The number of aboriginal people migrating to cities has increased since 1961, from 200 in that year surging up to a peak of 11,360 in 1994. This figure accounts for 32.4% of the total aboriginal population (Ministry of Interior, 1995, 6). Nonetheless, in a relatively short period of 6 years from 1971 to 1976, the number of urban aborigines increased eight times, up to one-twelfth of the total aboriginal population (Li, 1982, 401-403).

The Mass Urban Migration of the Aborigines from the 1970s to the 1980s

The 1970s marked the critical period for the mass migration of the mountain aborigines to urban areas and their incorporation into Taiwan's booming economy as wage labour. The main force driving the aborigines to urban migration came from the social and economic developments in the late 1960s. The job opportunities created by the booming industrial areas around the cities exerted a strong pull on the mountain dwelling aborigines. This, accompanied by the push from the decline of agriculture in their original settlements, can explain the high visibility of the indigenous people in the urban territories since the late 1960s. According to a survey of northern aborigines, the majority of those who have now settled down in the greater Taipei area were originally the Ami, who migrated northward from Hualian and Taitung. They claimed 94% of the total population of urban aborigines in the Greater Taipei area. The fact that the Ami group became the largest tribe of migration reflects its plains-dwelling character. The generation-long experience of settling in the plains areas under a policy of no land protection had prepared the indigenous Ami people for the necessary skills to catch every opportunity in the process of urban expansion and growth to improve their living standards (Lin, 1983, 5).

On the whole, four types of towns drew the immigration of the aborigines. Destinations of the first type are the exterior counties surrounding the Greater Taipei area, such as Panchiao, Chungho, Hsinchuang, Hsintian, and Hsichi. The second type of destinations are mainly the industrialized towns near their original settlements, such as Pate, Pingcheng, Tahsi, and Taiyuan. The inland cities such as trade and industrial centres constitute the third type of destination for aboriginal urban migration. Migration destinations of the last type are mining and fishing towns, such as Keelung and Kaohsiung (Kuo, 1997, 51). This pattern fully reflects the fact that the trajectories of aboriginal urban migration are closely linked to the job openings of these destinations.

However, this prospect seemed to change in the 1980s. The descendants from other mountain tribes also joined this trend of migration to urban regions. According to the 1994 aboriginal survey, the percentage of tribal distribution among urban aborigines has changed greatly. The Ami tribe still accounts for the largest proportion of urban aboriginal population, but the percentage has fallen to 70%, followed by the Atayal which claim 17% and the Paiwan with 6% (Chen and Sun, 1994, 54).

With reference to the results of the two latest surveys on the living conditions of the urban aborigines, the aboriginal migrants to urban areas in the 1990s mostly came from the tribes of Bunun, Atayal, and Tsou. More interestingly, the group of aboriginal migrants who held the most positive view of urban life within the nine indigenous tribes was neither the Ami, who comprised the largest number of urban migrants, nor the Paiwan who were the second largest. The percentage of the population within these two tribes expressing a positive approach toward urban migration remained rather steady, roughly at the level of 35% between 1994 and 1995. By contrast, the tribal people originating from either the plains or mountain villages, especially the Rukai from Pingtung and the Tsou from Mountain Ali, showed the largest and second largest percentage of members expressing the most positive view of migration to the urban region. The aborigines from the Saisiat tribe represented another exceptional case, which had 66% of its total population holding a positive view toward urban migration and no big fluctuation during the two years of 1994 and 1995.

The Main Labour Markets of the Urban Aborigines

Within Taiwan's labour markets, aborigines seemed to have gradually assumed the role of unskilled workers, or coolies. Most aborigines were less educated and generally short of the specialist skills needed to take up urban jobs. Upon moving into the cities, they were usually recruited into unskilled jobs, which offer lower social status and poor wages. Machine operators, construction workers, mould workers, sea fishers, heavy goods carriers, and truck drivers— in general, the physically exhausting, highly dangerous, and poorly paid jobs —all were the occupations the aboriginal urban migrants relied on to earn their living. These types of jobs mostly lacked any potential for further promotion and attracted no local Taiwanese workers. Most of the jobs were temporary in nature, short of fringe benefits and job security. More often than not the aborigines in these jobs were requested to move from one site to another.

A majority of urban aborigines were concentrated in the manufacturing sector and the construction industry. In other words, the aboriginal urban migrants either became factory workers or offered their labour at construction sites. Apart from these, coal miners and fishers were the two major occupations with strong community connections into which the remaining aboriginal urban migrants were drawn. In the 1970s, aborigines migrating to the city of Kaohsiung were mostly engaged in the fishing industry and worked as sea fishers. The job is physically demanding and poorly paid.

> *The reward system adopted in the sea fishing industry was built upon the rule of bonus sharing. Most of the aboriginal urban migrants were employed to conduct near-ocean single- or double-netted fishing on boats of 50 or 100 tons. The base wage level was as low as NT$400 a month and remained unchanged during the past 10 years. In order to recruit a sufficient number of the workforce and gain their commitment to the jobs on the sea, the employers in the fishery industry prepaid each sea fisherman a lump sum of NT$20,000 as the subsistence fee for his family because one journey of net fishing would usually take the fishermen away*

> *from their family for a period of one to two months. Added to the base wage level, each of the fishermen would be awarded a NT$35 to 45 bonus from every net profit of NT$10,000. That is to say, if there was a NT$10,000,000 net profit, then each fisherman would gain NT$35,000 to 45,000 for one journey of two months. This did not seem too bad a deal, but the problem was that the employers seldom honestly reported the profits, so the monthly income of the fishermen usually fell short of NT$5,000. As a result, after two or three runs out for ocean fishing, not only did the fishermen not gain any profits, but got into a debt of NT$70,000 to 80,000 unpaid to their employers* (Lin, 1983, 18).

In the early 1980s, traditional agriculture was stricken by a further decline and the economic developments created by export-oriented industrialization widened the discrepancy between rural and urban districts. This led to a second wave of aboriginal urban migration. In contrast to the predominance of the Ami tribe among the aboriginal urban migrants in the 1970s, a majority of the aboriginal urban migrants at this time came from the tribes of Atayal, Paiwan, and Bunun. Because of the decline of the coal-mining industry, most of the second-wave aboriginal urban migrants were drawn into the flourishing construction industry.

Meanwhile, the boom of the construction industry led aboriginal urban migrants to develop a pattern of quasi-tribal employment practices in adaptation to the strangeness of their urban situation. For example, a large majority of the migrants from the Paiwan tribe to the Greater Taipei area spontaneously organized a work team to look for jobs in the informal sector of the construction industry (Yang Shi-fan, 1997). This pattern of the quasi-tribal employment practice not only brought the aboriginal migrants a feeling of security but also helped them earn higher wages. In a recent field study, one urban aboriginal interviewee described their work experience in subcontracting:

> *When we got familiarized with the skills used and made several contacts with various employers, we formed a work team to liaise about subcontracts. The profits of subcontracting were far greater. All our team members were unmarried men from the same village. Between us there is a close relationship of kin: someone is either the other's brother or nephew or cousin. The channels we lean on to get subcontract jobs are through the medium of friends, either our kind or newly-acquainted Han Chinese. We make friends with the foremen and even with the bosses in every place we get job offers. Once we are without work at hand, we contact these friends in different work sites* (urban aboriginal field notes file number 068).

The aboriginal experience of urban migration also involves a gender difference. The above account was the part of the story relating to young male aboriginal migrants. The adventure of female aboriginal urban migrants tells another story. Most unmarried female aboriginal urban migrants were either absorbed by the manufacturing industry as factory workers, or recruited by part of the labour-intensive service sector such as hotels, restaurants, and hairdressing salons. Undoubtedly none of these were well-paid jobs. In a field study tackling the social adjustment problems of aboriginal female urban migrants, Huang (1992) revealed that many hairdressing salons in the Greater Taipei area "bought off" aboriginal teenage girls from their parents through tribal brokers with a lump sum of about NT$ 30,000 (less than US$ 1,000) for a 2-year contract. In fact, some other studies also found that the common practice of the job offices in the mountain counties was the use of an unbelievably low level of wage rates to buy off aboriginal teenage girls from their parents and then sell them to local manufacturing factories as child workers. And tragically, some of them were even sold to be child prostitutes.

Generally speaking, the employment prospects for married and older indigenous women were far fewer than their unmarried aboriginal counterparts. Married aboriginal women migrated to cities with their husbands in an effort to earn a good living. Nevertheless, the discrimination against married women in the work place left little opportunity for them to be employed. Thus, most of them took up subcontracting work at home. Others came to the factories that hired their husbands for some part-time jobs, or followed their husbands or sons to the construction sites for occasional chores, or worked temporarily in the fish market (Huang, 1992, 206-209).

The Impact of Globalization: Competition from the Foreign Workers

The general picture of urban aboriginal migrants in the golden period from the late 1970s to the early 1980s was for the most part one of being integrated into some of the mainstream Taiwanese economy. Nevertheless, the period after the 1987 lifting of martial law was to the urban aborigines a period of marginalization, though it was to some extent marked by rapid expansion and economic restructuring. Despite the growth in jobs, unemployment rose substantially in the urban aboriginal population as increasing numbers of foreign workers entered Taiwan's labour markets. Several factors worked together to explain this unexpected situation. The most critical factor came from the 1989 legal change which allowed local employers to recruit foreign workers to resolve the problem of labour scarcity created by economic expansion.

Parallel with rapid economic growth was the transformation of Taiwan's employment structures in the late 1980s. In the 3 years to 1989, employment in the industrial sector and the service sector increased by 58,000 and 337,00 workers respectively. Significantly, in 1988, total service sector employment overtook that in the industrial sector for the first time. Between 1987 and 1989, a net 236,000 people entered the labour market. The breakdown for this figure was 161,000 leaving the agricultural sector, 60,000 entering the industrial sector, and 337,000 joining the service sector. The average annual growth rate of employment in the industrial sector during this period was about 0.6%, and a considerable portion of this increased labour force was made up of illegal foreign workers, especially from Southeast Asia.

Labour shortages in the manufacturing and construction sectors had been considered the main cause of soaring labour costs, the most serious obstacle to maintenance of Taiwan's labour-intensive export development. Some figures for labour shortages from the Statistics Department in the Executive Yuan (January 1989) reported vacancies of 265,000 jobs in the industrial sector and 56,000 jobs in the service sector. It was said that the real situation might be much more devastating in medium and small enterprises, the backbone of Taiwan's economy. From mid-1986 on, the number of low-skilled illegal foreign workers from Southeast Asia coming into Taiwan's construction and manufacturing sectors was officially put at around 10,000 to 30,000. Nevertheless, media estimates ran as high as 100,000 to 300,000. In order to rectify the illegal employment of a large number of foreign workers, the Council of Labour Affairs (CLA) finally gave foreign workers legal entry to Taiwan's job markets.

The introduction of foreign workers relieved the severity of the labour scarcity. In 1993 the number of vacancies in the eight major industries fell to 129,886—a reduction of 46%. In all the sectors of industry most afflicted by the problem of labour shortage, the number of vacancies in manufacturing and construction—the two aborigine-dominated sectors—was reduced respectively from 164,286 and 31,059 in 1989 to 86,783 and 6,140 in 1993. According to the latest official statistics (CLA, 1999), at the end of 1998, there were 263,747 foreign workers in Taiwan (Table 5.1). A majority of foreign workers were from Thailand with a total of 141,230 making up 50% of the foreign work force. The Philippine foreign workers (111,250) were ranked second, claiming 42%.

Table 5.1 Numbers of foreign workers from Southeast Asia

		Thailand		*Philippines*		*Indonesia*		*Malaysia*	
Year	*Total*	Persons	%	Persons	%	Persons	%	Persons	%
1994	151,989	105,152	69.2	38,473	25.3	6,020	4.0	2,344	1.5
1995	189,051	126,903	67.1	54,647	28.9	5,430	2.9	2,071	1.1
1996	236,555	141,230	59.7	83,630	35.4	10,206	4.3	1,489	0.6
1997	248,396	132,717	53.4	100,295	40.4	14,648	5.9	736	0.3
1998	263,747	132,257	50.1	111,250	42.2	19,466	7.4	774	0.3

Notes: 1. No statistics on foreign workers before 1994.
2. The figures were accumulated.

Source: Department of Statistics, Executive Yuan, 1998.

Worthy of specific attention is the fact that by the end of 1998 the total number of foreign workers had well outnumbered the total male aboriginal population currently employed in Taiwan's labour market. The average wage rate of most foreign workers was about two-thirds that of local workers and they endured long working hours, often more than 10 hours a day (Table 5.2). Pulling all the threads together, the tendency of cheap foreign workers to rob the urban aborigines of job opportunities in the sectors of manufacturing and construction has become ever more crystallized. According to the data provided by an aborigine-elected legislator (Kao, 1996), after the entry of foreign workers to Taiwan's job markets, the average monthly working days decreased from 25 days to 12 days. This apparently has been considered to be a factor fostering a deterioration of the living standards among urban aborigines.

Table 5.2 Comparisons of wages and work hours of local Taiwanese workers and foreign labourers

Items	*1993*	*1994*	*1995*	*1996*	*1997*
1. Wages (unit: ratio = wages of foreign workers divided by those of local workers)					
Total	84.83	86.21	88.78	93.50	96.01
Manufacturing	88.22	89.27	90.66	94.19	96.88
Construction	75.50	70.14	74.96	81.66	80.75
2. Work Hours (unit: hours = work hours of foreign workers minus those of local workers)					
	a. MANUFACTURING				
Average	27.73	28.30	31.22	31.82	36.76
Normal	3.88	10.82	8.91	11.90	11.70
Overtime	23.85	17.47	22.31	19.92	25.06
	b. CONSTRUCTION				
Average	44.29	34.35	42.15	45.86	58.42
Normal	9.12	12.84	15.19	12.44	18.67
Overtime	35.16	21.62	26.96	33.42	39.75

Source: *Survey Reports on Management and Utilization of Foreign Workers*, Council of Labour Affairs, 1998.

The employers' search for abundant cheap labour to attack wages and working conditions can be found even where there is no racial or ethnic difference. But working-class hostility toward immigrant workers is also universal. Nonetheless, in the recent context of Taiwanese society, the cost-cutting motivation of employers offered a unique opportunity for the urban aboriginal workers and local Taiwanese workers to form a united front in which to dissolve the misunderstanding and separation between them in terms of ascriptive factors. In addition, the little practical effort made by the CLA officials to incorporate urban aborigines into Taiwan's labour force in the 1990s, despite rhetoric on participation and equal opportunity, also fostered formation of a new alliance between workers from these two ethnic groups. Why was the effort not considered worthwhile by the government? The reason was rather paradoxical. The large utilization of urban aborigines in the manual jobs in construction and manufacturing could be regarded as a way to recognize the specialized skills of the urban aborigines in these sectors, but this temporary strategy might constitute institutional discrimination against the aborigines. In other words, the consequent long-term political and social costs would be too high to bear. Needless to say, neither was it considered to be worthwhile to employers. The urban aboriginal population was too expensive to make a major contribution to meeting large-scale labour demands. Hence the main official responses to the disadvantageous situation of the urban aboriginal population in the labour markets continued to be welfare-oriented, even though a nominal official policy of aboriginal employment promotion was promulgated as early as 1994.

In fact, public concern has loomed large over the loss of job opportunities of local people, especially the urban aborigines, caused by the introduction of foreign workers. Therefore increased pressure was built up to push CLA officials to adopt rigid measures to curb entry numbers for new foreign workers. Numerous attempts were made to push the CLA in this direction. Take the most recent case as an example: On March 3, 1998, Chairman of the Workers' Party, Luo Mei-wen, mobilized a group of 30 aboriginal leaders to express their discontent in front of the Council of Labour Affairs. They also protested the failure of government to protect the employment rights of indigenous people by overturning a table on which the pamphlets and flyers of official assistance for unemployed aborigines were displayed. This attack doubtless drew a positive but transient response from CLA officials. Regretfully it only earned a nominal agreement for the activists with a cut in the quotas of imported foreign workers and an adoption of a case-by-case review measure in deciding whether official permission would be granted to major government construction projects estimated at over NT$200 million (US$5.8 million). Added to this, CLA officials also promised that after the passage of the transitional period, no permission would be granted to employ foreign workers in manufacturing firms with low economic value, little prospect of development, or high pollution.

Taiwan's unemployment rate increased on average to 2.93% of the labour force, amounting to 280,000 people in the late 1990s (the unemployment rate was highest at 3.19% in 1996), while the rate of urban aboriginal unemployment reached 4.12%. This has caused the emergence of a counter current among the urban aborigines returning to their home counties in anticipation of being able to get a job there. In spite of the fact that Taiwan's unemployment rate was running high, employers of foreign workers in the service sector were complaining that a 3-year turnover rate of foreign workers had hindered their competitiveness. (Note: The employment of foreign workers was usually based on a 2-year fixed-term contract and only a 1-year extension was granted after their original contract expired. At the end of 3 years, the foreign worker had to leave the island and could not return for work purposes.) For this reason, the Service Employers' Association actively lobbied for permission to re-import workers. In addition, foreign workers developed an organized front to voice a desire to work in Taiwan beyond the current 3-year limit.

In response, an official survey (Directorate General of Budget, 1998) was conducted and its findings indicated that 30% of the currently unemployed had little inclination to fill in the vacancies in the low-paid menial job markets (Table 5.3). By implication, this survey seemed to provide a reasonable excuse for the claim that more foreign labourers would still be needed to maintain continued domestic economic growth. Therefore, despite great opposition from aboriginal groups and labour unions, the CLA still tried to revise the foreign workers' work permit policy with the aim of extending the maximum 3-year stay to 9 years. In recent talks initiated by service-industry employers, the CLA proposed a measure of work permit extension that would allow foreign workers to apply for extensions up to three times and would grant them a maximum stay of 9 years on the island.

Table 5.3 Vacancies in construction and manufacturing

	1990	*1992*	*1993*	*1994*	*1995*	*1996*	*1997*	*1998*
Total	-	-	-	-	233,707	200,633	193,555	196,450
Percentage (%)	7.10	8.16	8.10	4.08	4.00	3.52	3.38	3.42
Industry	-	-	-	-	127,252	114,038	116,966	115,247
Percentage (%)	8.81	8.06	7.19	3.73	4.11	3.81	3.89*	3.83*
Manufacturing	-	-	-	-	109,240	97,553	101,562	99,517
Percentage (%)	8.75	8.63	7.79	4.07	4.35	3.99	4.10	4.01
Construction	-	-	-	-	15,310	15,060	15,282	15,361
Percentage (%)	11.53	5.26	5.06	2.32	2.86	3.04	3.15	3.23
Tertiary	-	-	-	-	106,455	86,595	76,587	81,418
Percentage (%)	3.63	8.29	8.17	4.62	3.88	3.21	2.82*	2.97*

Notes:
1. No corrected figures are available in 1990-1994.
2. The vacancy rate = vacancies / vacancies + the employed population
3. * stands for estimates.
4. The 1998 figure is based on the data up to September of 1998.

Source: Directorate-General of Budget, Accounting and Statistics, Executive Yuan, *Survey of Monthly Earnings of Employees, 1990-September 1998.*

Marginalization and Exclusion

The concept of an under class in Britain was developed in the 1970s to show that ethnic minorities with various immigration backgrounds tend to be severely disadvantaged in terms of opportunities for economic advancement. Compared to whites, non-whites had a different kind of position in the labour market, a different housing situation, and a different form of schooling (Rex and Tomlinson, 1979). In the same vein, this concept in the US describes groups permanently trapped in poverty. In Marxist terminology, the equivalent term is a sub-proletariat. The sub-proletariat argument suggests that discrimination and marginal status led immigrant minorities to over-representation among the socially and economically disenfranchised (Sivanandan, 1982; Castles and Kosak, 1985).

It must be admitted that the applicability of an under class or a sub-proletariat concept to describe the disadvantageous economic and social conditions of immigrant ethnic groups in Britain and the US remains a controversial issue. However, the concept of a sub-proletariat provides a useful conceptual tool to approach the urban experience of the aboriginal migrants in the context of Taiwanese society after the introduction of imported foreign workers. The first-wave aboriginal migrants in the late 1970s might have found reasonably acceptable explanations, such as lower education, poor literacy, and strangeness for their unfavourable positions in the job markets. However, as more and more job opportunities went to poorly-educated southeastern Asian foreign workers, the feeling of deprivation strengthened their sense of inferiority.

As well as wages and labour costs, the quality and levels of skill are other criteria adopted by the employers to decide whom to employ. In the economic booms of the late 1970s and early 1980s, aboriginal urban migration was mainly a result of economic considerations—a search for better job opportunities. Aboriginal urban migrants went where expanding industries needed them most, and this had been overwhelmingly to the industrial suburbs around the big cities or to expanding towns reliant on specific industries. The concentration into certain geographical areas was matched by segmentation in employment experiences of urban aborigines—poor working conditions, unhealthy working environments, high industrial accident rates and illness, and so on. Although the concentration of aboriginal urban migrants in low status jobs affected their material well being and social prestige, many local construction employers came to value the skills of urban aborigines (Lin Chao-chen, 1993). Nevertheless, since the backbone of Taiwan's economy consisted of numerous small and medium enterprises, in the eyes of employers the importance of cutting labour costs usually outweighed the importance of employing workers with good skills. After the introduction of foreign workers gained official support, we saw some of the highest aboriginal unemployment rates spread across various geographical areas. The southeastern Asian foreign workers might have provided a much-needed addition to Taiwan's overstretched labour force.

However, it is also a fact that their coming has not only undermined the value of specialized skills possessed by urban aboriginal workers, but has also given rise to a new pattern of discrimination. When there was no longer full employment and unemployment figures climbed, the dark foreboding became crystal-clear—it not only aroused the xenophobia of urban aborigines against foreign workers but also elevated the tension between the aborigines and local Han Chinese. This aboriginal discontent has already worked two ways. On the one hand, it increased the tension between the urban aborigines and the Han Chinese and worked negatively against the formation of an island-wide nationalism. On the other hand, the discontent has already been used by aboriginal activists to promote some deep reflection on the secondary status in citizenship experienced by the aborigines. That might positively contribute to the emergence of a pan-aboriginal consciousness much needed in pursuit of an affirmation of their equal rights as citizens.

To summarize, one cannot be optimistic that the integration of urban aborigines is an easy task. Increasing global economic integration has highlighted the disadvantageous position of urban aborigines in the face of imported south-eastern Asian workers. The unpleasant and lamentable encounters examined in this chapter are by no means made-up stories, but real experiences of many urban aborigines with solid support in the findings of many studies. In the less inhibited political environment of today, it is expected that the discovery of ethnicity as a form of social mobilization and identification by aboriginal activists might replace the old mainlander-Taiwanese bifurcation within the big ethnic grouping of Han Chinese in the foreseeable future of Taiwan's politics. All in all, what one would hope this chapter would achieve is to promote an understanding of the problems of urban adaptation and employment conditions faced by aboriginal migrants. From the viewpoint of social science, these problems of ethnic assimilation and conflict are now of central importance. From the perspective of an ordinary citizen they are, along with economic growth, the central problem of politics.

References

Bodley, J.H. (1982). *Victims of progress*, 2nd ed., Palo Alto, CA: Mayfield Publishing Company.

Bureau of Labour, Taiwan Province Government (1991). *Taiwan sheng shan bao yu chuan chuan yuan sheng huo zhuang kuang diao cha bao gao, Taiwansheng laogongchu.*

Bureau of Statistics, Department of Interior (1995). *Zhong hua min guo ba shi san nian Taiwan di qu du shi yuan zhu min shenghuo zhuang kuang diao cha bao gao*, Taipei.

Bureau of Statistics, Department of Interior (1996). *Zhong hua min guo ba shi si nian Taiwan di qu du shi yuan zhu min sheng huo zhuang kuang diao cha bao gao*, Taipei.

Cowlishaw, G. (1988). *Black, white or brindle: Race in rural Australia*. Cambridge: Cambridge University Press.

Council of Labor Affairs, Executive Yuan, *Wai ji lao gong de guan li yu yun yong diao cha bao gao,* Taipei: Executive Yuan.

Fu, Yang-zhi (1985a). Du shi shan bao yan jiu de hui gu yu qian zhan, *Si yu yan*, 23(2), 65-81.

Fu, Yang-zhi (1985b). Du shi she hui de te zhi yu yi min yan jiu, *Si yu yan*, 23(3), 321-343.

Fu, Yang-zhi (1993). Du shi shan bao Ami zu de ju ju sheng huo xing tai: yi xi mei she qu wei li, *Bulletin of the Institute of Ethnology Academia Sinica*, 74, 163- 214.

Gastles, S., and Kosack, G. (1985). *Immigrant workers and class structure in Western Europe.* London: Oxford University Press.

Gao, Yang-sheng (1997). Cong yuan zhu min jiao du si kao she fu zhi du: ta men de wen ti bu shi she hui wen ti, er shi zhong zu wen ti, *Zhong yang ri bao*, 3(November), 7.

Hannerz, U. (1974). Ethnicity and opportunity in urban America. In A. Cohen (Ed.), *Urban Ethnicity* (pp. 37-76). London: Tavistock.

Huang, Mei-ying (undated). Du shi shan bao yu du shi ren lei xue: Taiwan tu zhu zu qun du shi yi min de chu bu tan tao, *Si yu yan*, 23(2), 294- 219.

Jenkins, R. (1986). Social anthropological models of inter-ethnic relations. In J. Rex and D. Mason (Eds.), *Theories of race and ethnic relations* (pp. 170-186). New York: Cambridge University Press.

Li, Ming-zheng (1997). Da tai bei di qu du shi yuan zhu min sheng huo mo shi ji qi bian qian kuang de bi jiao fen xi, *Taiwan yuan zhu min li shi wen hua xue shu yan tao hui lun wen ji,* Nantou: Taiwan sheng wen xian hui.

Lin Jin-pao (1981a). 'Bei qu Du shi shan bao Sheng huo zhuang kuang Diao cha yan jiu', Nan tou: Taiwan sheng zheng fu Min zheng ting.

Lin Jin-pao (1981b). Taiwan bei bu de du shi shan bao, *Zhong guo lun tan*, 12(7), 24-28.

Lin Jin-pao (1983). 'Taipei yu Kaohsiung shan bao ju min sheng huo zhuang kuang diao cha,' Nantou: Taiwan sheng zheng fu.

Moody, R. (1988). *The indigenous voice: Visions and realities*. London; Atlantic Highlands, NJ: Zed Books.

Nan dao shi bao:

1995/07/13:5 'Wai lao qiang zou fan wan, yuan zhu min cheng le du shi you mu min zu.'

1995/07/27:1 'Yuan zhu min si nian ji hua dui kang wai lao yin jin zheng ce: Sheng fu meng xia "Xin pin jiu jiu" yao fang, li wan kuang lan mian lin kao yan.'

1995/09/08:4 'Zhong shi lao gong an quan, fang zhi zhi ye zhai hai: gao wei xian de jian zhu gong di dou you yuan zhu min rong gong de zu ji.'

1995/12/29:4 'Taiwan di qu du shi yuan zhu min sheng huo zhuang kuang diao cha, xian shi du shi yuan zhu min zui qi pan chuang ye dai kuan.'

1996/01/05:2 'Wai lao lao guo jie, yuan zhu min bei hei guo.'

1996/03/12:4 'Qiu zhi bu yi, zhuan ye ye nan, jing ji bu jing qi: yuan zhu min qian tu mang mang.'

1996/06/13:4 'Wei le yuan zhu min ge ai wai lao : sheng lao gong chu' lao wei hui chong xin ping gu wai lao zheng ce, dui yuan zhu min duiyu ye fu dao jiang you xian jin xing, ti gao jiu ye liu.'

1996/10/17:5 'Shi ye ren kou yuan zhu min zhan shou wei: lao wei hui ren wei yuan zhu min ying gai can jia zhi xun ti sheng ji shu neng li.'

1996/11/02:4 'Du shi zhong de yuan zhu min xi lie bao dao 1-4.'

1996/11/07: 2 'Zheng fu tu li wai lao, yuan zhu min mei tou lu.'

1996/11/07: 6 'Shi ye wei ji zhong de du shi yuan zhu min feng nian ji: yuan zhu min zhi lao gong lian meng yu qing zheng fu tiao zheng zheng ce da kai yuan zhu min sheng cun kong jian.'

1996/11/14: 3 'Yuan zhu min lao gong da sheng na han: xiang zheng fu tao yi wan fan chi.'

Rex, J., and Tomlinson, S. (1979). *Colonial immigrants in a British city: A class analysis*. London: Routledge & Kegan Paul.

Sivanandan, A. (1982). *A different hunger: Writings on Black resistance*. London: Pluto Press.

Werther, Guntram F. A. (1992). *Self-determination in Western democracies: Aboriginal politics in a comparative perspective*. Westport, Conn.: Greenwood Press.

Xie, Gao-qiao (1993). Wai ji lao gong yu yuan zhu min jiu ye wen ti, *Jiu ye yu xun lian*, 11(5), 10-12.

Xie, Gao-qiao (1995). Taiwan da du hui yuan zhu min de ju zhu ge li - yi Taipei shi wei li. In Cao Jun-han yu Ke Qiong-fang (Eds.), *Zhong xi du hui qu zhi fa zhan yu mian lin de wen ti*. Taipei: Zhong yang yan jiu yuan ou mei yan jiu suo.

Xie Gao-qiao et al. (1991). *Taiwan shan bao qian yi du shi hou shi ying wen ti zhi yan jiu*, Taipei: Xing zheng yuan yan jiu fa zhan kao he wei yuan hui.

Yang, shi-fan (undated). 'Paiwan zu cheng xiang qian yi zhe chuan tong wen hua yu she hui zu zhi zhi chi xu yu zhuan bian - yi ping he cun tai wan bei bu yi min qun wei li', unpublished Masters Thesis in Sociology.

Zhan, Huo-sheng et al. (1991). Taipei di qu du shi shan bao qing nian sheng huo zhuang kuang yu jiu ye xu qiu zhi yan jiu, Taipei: xing zheng yuan qing nian fu dao wei yuan hui.

Zhang, Qing-fu (1993). Yuan zhu min de jiu ye wen ti, *Jiu ye yu xun lian*,11(2), 7-9.

Zhang, Xiao-chun (1974a). Taipei di qu shan di yi min shi ying chu bu diao cha yan jiu: I, *Si yu yan*, 6(11), 1-21.

Zhang, Xiao-chun (1974b). Taipei di qu shan di yi min shi ying chu bu diao cha yan jiu: II, *Si yu yan*, 12(1), 30-37.

Zhong guo shi bao China Times

1995/09/11: 17 'Ren kou fan zi pian shang chuan, tao jin zhi lyu bian le diao- zhong jie he yue you ru mai sheng qi'yuan zhu min lao gong yan zhong shou bo xue.'

1995/08/14:7 'Wai ji lao gong qiang fan wan - yuan zhu min jiu ye la jing bao. Zhong shi wan bao '

1996/05/01:4 'Yuan zhu min yao qiu ting zhi yin jing wai lao: mei nian jian shao shi wan ming e.

1996/07/12:23 'Du shi bian chui de yuan zhu min wen hua yun dong: xi zhi hua dong xin cun de Ami zu ren.

Zi li wan bao

1996/05/05:4 'Zheng gong zuo quan , yuan zhu min lao gong jng zuo: lao wei hui da ying jiang yao qiu gu zhu you xian gu yong yuan zhu min cai ke huo wai lao pei e.'

Plate 7 A view of Chiang Kai-shek (Chiang Chieh-shih) Memorial Park, Taipei ▶

Social Structure and Leadership: A Study of Organizational Commitment

Stanley T. Lee

Associate Professor, Institute of Interdisciplinary Studies, National Sun Yat-sen University

Objective of the Study

As often found in other sociological works, employment relations in this study are assumed to be embedded in the work structure of a complex bureaucratic organization. The author is interested in finding how work and organizational structures explain and/or specify the relationship between the leadership of supervisors and the work problems perceived and experienced by their subordinates, which are assumed to have degenerative effects on organizational commitment. Specifically, organizational divisions and positions are introduced to examine the relationship between leadership ability and attitude, and work problems in order to understand the nature of the relationship between supervisors and their subordinates. It is believed that the variation and the intensity of work problems reflect both the organizational structure and the leadership behaviour, and are often attributable to supervisors' behaviours in a work unit. The problems of work overload, work inequity, and role ambiguity are introduced into the analysis. Ultimately, the objective of this study is to understand the processes through which a supervisor generates organizational commitment from subordinates in complex organizations. Further theoretical discussion focusing on the comparison of the patterning behaviour between Taiwan and other capitalist societies is also attempted. Although there are no data available for direct comparison, it would still be interesting to know, perhaps through a review of literature and scientific reference, whether the relationship between a leader's ability and attitude, and work problems show any unique Taiwanese characteristics.

The Data

The data of 1,299 employees were selected out of an original sample of 1,357 cases collected by a well-designed mass survey questionnaire administered to randomly selected employees of a previously state-owned Taiwanese steel company in 1990 (Lee and Tu, 1991; Young et al., 1991). Those who did not provide complete or adequate information on the items measuring organizational commitment behaviours were eliminated from the analysis.

The steel company was one of 712 state-owned enterprises in 1990. It was officially established in 1971, but not in actual operation until 1978. By 1990 the company had almost 10,000 employees. Among them were 1,584 professionals, 1,263 administrators or supervisors, and 6,785 general and non-operational workers. Their positions reflect various educational backgrounds and were distributed in eight major organizational divisions: administration, business, finance, aluminum production, technology, steel production, research and development and engineering, and secretary/industrial engineering. Due to its nature as a steel company, and its youth in terms of number of years since its establishment, almost all of its employees were men between 30 and 40 years of age. There were only 189 female workers, mainly in the division of administration. The sample collected validly reflects the structure of the study population (Table 6.1).

The Problem and its Theoretical Considerations

Unlike the majority of well known Taiwanese private enterprises, which are not only small in size but also simple in structure, this company, as well as other state-owned enterprises, usually set up various functional departments with formal systems of regulation, hierarchy of positions and authority, and standardized working procedures (Hamilton and Kao, 1990; Hwang, 1990).

Different working units within these functional departments would possibly display various leadership behaviours and work problems, and their respective relationships. It is also believed that organizational commitment behaviours are a consequence of leadership behaviour and work problems. Leadership behaviours in various forms of organizational setting have been linked, at least partially, to role stress (Dellva et al., 1985), role conflict (Dellva et al., 1985), work equity (Schnake et al., 1995), role ambiguity (Jurma, 1978), job satisfaction (Schnake et al., 1995; Marsh and Atherton, 1981; Jurma, 1978; Orpen and Hall, 1994), job performance (Hunt et al., 1978; Rao et al., 1987; Orpen and Hall, 1994), interpersonal attachment (Yoon et al., 1994), and community life (Leviatan, 1994), in addition to organizational commitment (Glisson, 1986; Sharmir, 1991; Hopfl, 1992; Li et al., 1997; Minturn, 1995; Schnake et al., 1995). Among the latter, those relevant to task characteristics such as role stress, work equity, role ambiguity, and role conflict are, in some studies, considered to be leadership substitute variables including characteristics of subordinates and organizational characteristics (Podsakoff et al., 1993). These substitutes are believed to be someone or something in the leaders' environment that reduces their ability to influence subordinate attitudes, perceptions, or behaviours. They are considered to be able to replace, in effect, the impact of leaders' behaviours

Table 6.1 Comparison between sample and population structures

	Sample		*Population*	
Characteristics	N	%	N	%
Gender				
Male	1,334	98.3	9,553	98.1
Female	23	1.7	189	1.9
Age				
Under 30	176	12.9	1,514	15.5
30 – 34	335	24.7	2,501	25.7
35 – 39	532	39.1	3,840	39.4
40 – 44	245	18.0	1,319	13.5
45 – 49	49	3.6	400	4.1
50 and over	22	1.5	1,065	1.7
Education				
Graduate degrees	32	2.4	218	2.2
University	269	19.8	1,863	19.1
College	221	16.2	714	7.3
Senior/vocational high	722	53.0	5,653	58.0
Junior high	67	4.9	583	6.0
Elementary school	50	3.7	708	7.3
Position				
First rank administrators	11	0.8	73	0.8
Second rank administrators	41	3.0	254	2.6
Third rank administrators	46	3.4	292	3.0
Fourth rank administrators	99	7.2	644	6.6
Professional/engineers	211	15.4	1,584	16.3
Non-operational workers	84	6.1	419	5.3
Operational workers	877	64.1	6,366	65.4
Division				
Administration	45	3.4	297	3.1
Business	67	5.0	481	5.0
Finance	28	2.1	188	2.0
Aluminum production	54	4.1	411	4.3
Technical	106	8.0	841	8.7
Steel production	950	72.0	6,956	72.2

(Podsakoff et al., 1993; Orpen and Hall, 1994). The extent and the way leaders exercise their power and control over their subordinates have been widely reported in the literature of organizational commitment (Jurma, 1978; Hunt et al., 1978; Gastil, 1994; Rao et al., 1987; Smagin, 1996). Earlier studies by Lippitt and White (1939) already indicated that different leadership styles have varying degrees of impact upon young people's behaviours with respect to their satisfaction, frustration, and aggression. Young people under authoritarian leadership are more likely to reveal aggressive or indifferent behaviours than are those under democratic leadership. The style of *laissez faire* (permissive leadership), however, results in the most aggressive behaviour among young people. Based on 1930s-1940s experiments conducted by US industrial sociologist Kurt Levine, Simagin (1996) found the styles of leadership all have significant advantages for Russian scientific reform, but still with varying team effects. Authoritarian methods achieve high results only in the initial phase of work, while democratic leadership achieves better results in terms of communication between team members and degree of work satisfaction. Permissive leadership is effective during the initial phase of research in basic sciences. Similarly, a study of 120 managers in India by Jaggi (1977) confirms a relationship between leadership style and job satisfaction. That is, job satisfaction is possible under consultative management, but not under authoritative management.

However, Gastil (1994) reported no direct relationship between leadership style and group productivity and members' satisfaction. According to Gastil, the relationship should take into account substitute factors, such as task complexity. Democratic leadership does appear to be positively associated with member satisfaction, particularly in simple and somewhat complex tasks. There is no relationship found in highly complex tasks. Gastil's study is a reflection of a contingency theory of leadership originally established by Fiedler (1972). In order to be effective in changing subordinate behaviours, leader behaviour must reflect the leader's status in the organization, the organizational structure, and the relationship between leader and subordinates. If such a reflection is achieved, leader behaviour in a work unit should be able to increase interpersonal attachment among subordinates. Furthermore, interpersonal attachment is supposed to have a positive effect on commitment to the work organization encompassing the work units, particularly when it occurs between dissimilar positions (Yoon et al., 1994).

The Hypotheses and the Study Variables

Organizational structure that promotes commitment to the group usually depends on certain mechanisms. Some of them are cultural and social insulation (Bittner, 1963; Simmons, 1964), keeping a distinctive language (Hostetler, 1988), specific dress (Isichel, 1964; Redkop, 1969), geographical isolation, (O'Dea, 1957), witnessing (Gerlach and Hine, 1970; Harrison, 1974; Shaffir, 1978), and various interpersonal influence tactics and strategies (Lofland, 1977, 1978; Snow, 1976; Snow and Phillips, 1980). Though leadership has been widely assumed to be a paramount factor in the study of organizational efficiency, few have specifically investigated leadership behaviour when searching for an explanation of the employment relationship between supervisors and their subordinates (Eisenberger et al., 1990; Howell and Dorfman, 1986; McElroy and Shrader, 1986). Are the work problems of subordinates, such as overload, inequity, and conflict, a result of their leaders' behaviours? And, would the impact of leaders' behaviour be adjusted to the nature of the social structure of both the work units and the entire complex organization encompassing these units?

The influence of a leader upon their subordinates can be attributed to the power of their position and their control over social resources in the organization, and the successful exercise of that power (Strasser, 1981; Portes, 1998). Thus, leader ability and attitude are hypothesized in this study to have significant and substantial impact upon organizational commitment with respect to work units and other organizational structures, but generally through the work problems perceived and experienced by the subordinates. More explicitly, the work problems experienced and reported by the employees of the steel company in this study, who voluntarily chose to participate, are limited to work overload, work inequity, and role ambiguity. Work overload reflects the burden and the difficulty of the assigned task. Work inequity is a psychological stress reaction to one's own task after comparing it to other's tasks in terms of burden and difficulty; while role ambiguity is also a psychological reaction, but specifically to the coordination between the leader and work units.

These work problems are believed to have degenerative effects on organizational commitment behaviours, and are also assumed to be a result of a leader's ability and attitude. It is also assumed that the variation of work problems in organizational divisions and positions is a reflection of leadership as shown in the leader's ability and attitude. Finally, organizational commitment in this study was measured by a 15 item scale developed by Mowday, Steers, and Porter (1979), with a minor modification (Lee and Tu, 1991). The ability of a leader was measured by a combination of indicators measuring skills in dealing with technical problems, organizational wit, planning and scheduling for work events, motivating subordinates, self expression, human relations, giving rewards to subordinates, and problem solving. In addition, the attitude of a leader was composed of indicators measuring personality traits and behavioural tendencies. Specifically, this variable was constructed to reflect understanding of subordinates' working situations, consideration of their difficulty in work, way of assigning tasks to subordinates in terms of fairness and justice, patience in listening to subordinates, and communication to subordinates. Measures are all based upon the perceptions of subordinates. They are supposed to be able to differentiate the scores measuring work overload, work inequity, and role ambiguity for all the divisions and positions of the organization.

THE FINDINGS

Organizational Commitment

The work problems perceived and reported by all employees, according to our data, are differentiable by organizational division, but not by organizational position. In general, those working in the division of administration were found to be relatively less likely to report the problems of work overload, work inequity, and role ambiguity than were those working in the divisions of production and technology (Table 6.2). The most likely to report these problems were those working in the division of technology. However, when administrators, professional engineers, and operational workers were compared, they did not show any variation in any one of these problems. The work problems under investigation in this study are very likely phenomena of organizational division rather than organizational position.

Table 6.2 Gamma score indicating the relationship between structural factor and work

Problem	*Work overload*	*Work inequity*	*Role ambiguity*
Position[1]	-.045	-.044	-.060
Division[2]	.113*	.188**	.180**

*p<.05 **p<.01

1. Position: administrators vs. engineers vs. workers
2. Division: administration vs. production vs. technology

These work problems were found, as expected, to be negatively associated with organizational commitment (Table 6.3). The greater the burden, inequity, and ambiguity the employees experienced in their work and role, the less likely they would be to commit themselves to the company. Commitment refers here to "the willingness of people to do what will help the group" (Kanter, 1978, p. 66). It means that they are willing to attach themselves to the requirements of their employment relationships with respect to their supervisors and co-workers. In another sense, when a person is committed to his/her group, s/he will gladly accept and do whatever task or work role is assigned. There is no role distance (Goffman, 1961). Such a commitment is nearly identical to the essence of Durkheim's "mechanical solidarity," Toennies' "*gemeinschaft*" system of social relations, and Simmel's "secret society." The committed individual is so completely tied to the group that s/he is totally dedicated to, and encompassed by, the functions of the group. As Kanter (1972) said, when a person is committed, "what he wants to do (through internal feeling) is the same as what he has to do (according to external demands), and thus he gives to the group what he needs to nourish his own sense of self" (p. 66-67). However, the pattern of work problems being related to organizational commitment is not, according to our data, a universal phenomenon. Among the three problems studied, work overload seems to be relatively more likely to draw a deviation from organizational commitment than are either work inequity or role ambiguity (Table 6.3). Work inequity appears to be the least likely to reduce a person's commitment.

Table 6.3 Gamma scores indicating the relationship between work problem and organizational commitment by organizational division and position

	Work overload	*Work inequity*	*Role ambiguity*
All employees	-.211**	-.079*	-.150**
Division			
Administration	-.275*	-.336**	-.338**
Production	-.223**	-.052	-.152**
Technology	-.087	.009	.029
Position			
Administrators	-.217*	-.036	-.031
Engineers	-.010	-.131	-.039
Workers	-.252**	-.081	-.202**

*p<.05 **p<.01

When organizational division was introduced, different patterns of the relationships surfaced specifically (Table 6.3). In the divisions of administration and production the relationships were strengthened, with one exception. Work inequity was no longer a degenerative factor of organizational commitment in production work units. However, in the division of technology every one of the relationships became negligible. Hence, work problems did not account for alienation behaviours of the employees working in the technology sector—not, at least, as shown in the company under investigation. Similarly, after introducing the factor of organizational position, the uniformity of the relationships between work problems and organizational commitment also disappeared. The relationships appeared to be strengthened for operational or similar level workers, but became negligible for both administrators and operational workers, with the exception of work inequity. Nevertheless, the relationship between work overload and organizational commitment remained as strong as it was for the administrators.

As such, work problems as seen in the work units of a complex organization are probably one of the reasons an employee would not want to commit to the organization. The way in which these problems emerge is rather complex, as the pattern of the relationships appears to be similar between the sector of technology work units and among professional engineers. In both cases, work problems show little impact on organizational commitment. There is also a similar pattern found among the sectors of production units and administration units and operational workers. In contrast to the former, however, their relationships are substantially strong. Thus, this difference suggests a possibility that the effect of work problems on organizational commitment is a reflection of the nature of the work units, and partially a function of the leadership behaviours of the supervisors located within these various work units.

The Effects of Leader Ability

The ability and attitude of supervisors were operationally and empirically constructed to make inferences about their behaviours and the impact they might have on the work problems and commitment behaviours of their subordinates. Their ability was directly derived from the opinions of subordinates. It consists of skills in organizing, scheming/planning, giving rewards, and problem solving as well as dealing with technical problems, expression/communication, and human relations. Their attitude was also compiled directly from the opinions of subordinates. It is a composite scale including attitude issues of understanding, fairness, justice, consideration, patience, and communication.

As can be seen in Table 6.4, the work problems of subordinates were found to be significantly and negatively associated with leader ability. To some extent, however, work overload and role ambiguity seemed to be more relevant to leader ability than work inequity; and, specifically, the skills in techniques, human relations, and problem solving were found to be relatively more important than other skills (Table 6.A1). Nevertheless, the association of work problems with leader ability remained substantial, even after organizational division and position were taken into account. Though it might not be significant in a statistical sense, the leadership ability of supervisors was found to be relatively more important for general and operational workers and those working in the production and administration sectors than for those in the technology sector. By looking into the situation experienced by administrators and professional engineers, the patterns were found to be somewhat contradictory to each other. The ability of administrators' supervisors was rather important in regard to the overload problem, but not to problems of work inequity and role ambiguity. On the other hand, the ability of professional engineers' supervisors was rather important if it was intended to explain work inequity and role ambiguity. It seemed not very useful for the problem of work overload. Supervisory ability is, indeed, at least partially responsible for work problems such as overload, inequity, and ambiguity as is shown in our data (Table 6.4).

Table 6.4 Gamma score indicating the relationship between leader ability and work problem by structural factor

	Work overload	*Work inequity*	*Role ambiguity*
All employees	-.243**	-.141**	-.199**
Division			
Administration	-.135	-.236*	-.209
Production	-.282**	-.119**	-.212**
Technology	-.147	-.159	-.133
Position			
Administrators	-.194	-.071	-.075
Engineers	-.119	-.293**	-.237**
Workers	-.283**	-.122**	-.217**

*p<.05 **p<.01

The influence of supervisory ability on work problems seems clearer for general workers, including both operational and non-operational employees, especially if they are working in the administration and production sectors. For engineers and professional workers, as well as those in the rank of administrators, the ability of their supervisors may have something to do with their work problems such as work inequity and role ambiguity, but certainly is not a factor militating against their commitment to the organization.

The Effects of Leader Attitude

As shown in Table 6.5, the attitude of supervisors is as important as their ability in explaining the variation in work problems. Attitude was found to be substantial and significantly associated with work problems such as overload, inequity, and ambiguity. Further analysis of the specific components of attitude revealed that the tendency to display understanding consistently stood out as the most important attitude variable (Table 6.A2). In contrast, communication attitude was unexpectedly found to be unable to explain any variation in work problems.

Nevertheless, attitude in general was also found to be more critical, relatively speaking, to the problems of work overload and role ambiguity than of work inequity as was the case for supervisory ability. Only a slight difference was detected in the patterning behaviour of the relationships between work problems of subordinates and the attitude of the supervisors, even after organizational division and position were introduced. The attitude of the supervisors in the sector of administration was found to be significantly associated with the problem of work overload alone, while their ability was found to be significantly relevant only for problem of work inequity.

A variation was also found among administrators. Their attitude was much more influential than their ability in the problem of work overload. Nevertheless, like the patterning behaviour of the ability of supervisors, their attitude was also found to be more relevant to the work problems of general workers and those working in production and administration sectors. It was relatively less relevant to those working in the technology sector. The situation experienced by administrators and professional engineers also displayed a contradictory impact as in the case of the impact of the ability of the supervisors. The attitude of the supervisors was relatively more relevant to the problem of work overload for administrators, but it was more relevant to the problems of work inequity and role ambiguity for professional engineers.

Thus, both the ability and the attitude of the supervisors are considered to be at least partially and equally responsible for the behaviours of their subordinates, especially, regarding to the work problems experienced and perceived by their subordinates. Although their behaviours, such as their ability and attitude, may account for the behaviours of their subordinates, especially, if they are general workers working in administration and production units, such a reference cannot be made about the behaviours of administrators and engineers including professional workers. The work problems experienced and perceived by administrators and professional engineers, in particular, within the administration and the technology sectors, are sometimes blamed on their supervisors. But very few of them would use these problems as an excuse for reducing their commitment to their group or organization.

Table 6.5 Gamma score indicating the relationship between leader attitude and work problem by structural factor

	Work overload	*Work inequity*	*Role ambiguity*
All employees	-.243**	-.140**	-.203**
Division			
Administration	-.297*	-.217	-.229
Production	-.238**	-.129**	-.214**
Technology	-.182	-.130	-.140
Position			
Administrators	-.261**	-.096	-.177
Engineers	-.156	-.276**	-.237*
Workers	-.250**	-.130**	-.209**

*p<.05 **p<.01

Discussion

The commitment of an individual member to his or her group or organization is often considered to be almost as important as role stress, role conflict, role ambiguity, job satisfaction, work turnover, and alienation. It is believed to be at least one of the keys to understanding the survival of the organization, including its development and decline (Whetten, 1987). In some studies, the behaviour was even extended to the issues of national competition in business and industry (Cole, 1979; Kalleberg, 1987).

Some sociological studies, on the other hand, continue to dig into the issues of bureaucracy, labour relations and processes, class conflict, and the other sociological phenomena, in order to search for an answer for the traditional question in social order and social control (Schwartz and Ogilvy, 1979; Lincoln and Laleberg, 1985). Actually, it is an issue of social relationships, socialization, and social integration if we can turn our attention to the relationship between the dimensions of organizational commitment and the supervisor-subordinate employment relations in a work setting (Mortimer and Lorence, 1979). From a practical methodological perspective, the elaboration of the relationship between leadership and the behaviour of organizational commitment may turn out to be a better explanation of why an individual wants to attach, become alienated, and eventually leave the group or organization (Adler and Adler, 1987). Such an answer is not only relevant to a person's identification, involvement, and dedication, but is definitely also related to the theoretical problems of socializing, maintaining relationships, and keeping an identity. Ferd Fiedler is one of the few who attempted to build a paradigm linking commitment behaviour to both leadership and organizational behaviour (Stogdill, 1974; Lincoln, 1989; Fiedler, 1967). This paradigm is generally based upon Weber's bureaucratic theory, but comes out with an emphasis on the success or failure of carrying out the authority by the leader of the organization. Since success or failure is very much dependent on (1) the leader-member relations, (2) the level of task structure, and (3) the power of the leader's position, it is also called contingency theory.

The ability and attitude of supervisors are certainly a reflection of the theoretically specified leader-member relations, as perceived by their subordinates in a large complex organization such as the Taiwanese state-owned company. It is no doubt that the supervisors are capable of functioning as change agents in their subordinates' behaviour, including the commitment of subordinates toward both their work units and the entire organization. However, the functions of the supervisors are probably specified by their positions in the hierarchical system and their work units, as Bums and Stalkes (1961) theorized. For example, a special kind of democratic leadership and autonomous management is probably required for the divisions of technology housing engineers and professional workers because it is more or less organic in its nature and relatively uncertain in its environment.

In contrast, an authoritative leadership and centralized management are probably more suitable for the divisions of production and administration where the majority of the labour force is skilled and semi-skilled, and their work is relatively more routine and stable (Hull, 1988; Green et al., 1996). Finally, although the work problem of the employees is likely a reflection of the leadership behaviours of the supervisors in an employment relationship, it is not a sufficient condition for the commitment behaviour under investigation. Work overload, work inequity, and role ambiguity are probably pernicious only in production work units, especially those working on assembly lines. Further analysis of the influence of leadership behaviour in organizational commitment is still needed, particularly for organic work units and professional workers.

References

Adler, P., and Adler, P. A. (1987). The social construction of organizational loyalty: A sociology of emotions model, an association paper of the American Sociological Association.

Bums, T., and Stalker, G. M. (1961). *The management of innovation*. London: Tavistock Publications, Ltd.

Cole, R. E. (1979). *Work, mobility, and participation*. Berkeley, CA: University of California Press.

Dellva, W. L., Teas, R. K., and McElroy, J. C. (1985). Leader behavior and subordinate role stress: A path analysis, *Journal of Political and Military Sociology*, 13(2), 183-193.

Eisenberger, R., Fasolo, P., and Davis-LaMastro, V. (1990). Perceived organizational support and employee diligence, commitment, and innovation, *Journal of Applied Psychology*, 75, 131-59.

Fiedler, F. (1967, 1972). *A theory of leadership effectiveness*. New York: McGraw-Hill.

Gastil, J. (1994). A meta-analytic review of the productivity and satisfaction of democratic and autocratic leadership, *Small Group Research*, 25(3), 384-410.

Glisson, C. C. (1986). The effect of leadership on workers in human service organizations, *Administration in Social Work*, 13(3-4), 99-116.

Glisson, C. C. (1988). Predictors of job satisfaction and organizational commitment in human service organizations, *Administrative Science Quarterly*, 33(1), 61-81.

Green, S. G., Aderson, S. E., and Shiver, S. L. (1996). Demographic and organizational influences on leader-member exchange and related work attitudes, *Organizational Behavior and Human Decision Processes*, 66(2), 203-214.

Hamilton, G. G., and Kao, C. (1990). The institutional foundations of Chinese business, *Comparative Social Research* , 12, 135-151

Hopfl, H. (1992). The making of the corporate acolyte: Some thoughts on Charismatic leadership and the reality of organizational commitment, *Journal of Management Studies*, 29(1), 23-33.

Howell, J. P., and Dorfman, P. W. (1986). Leadership and substitutes for leadership among professional and non professional workers, *Journal of Applied Behavioral Science*, 22(1), 29-46.

Hunt, J. G., Osburn, R. N., and Schuler, R. (1978). Relations of discretionary and nondiscretionary leadership to performance and satisfaction in a complex organization, *Human Relations*, 31(6), 507-523.

Hwang, K. (1990). Modernization of the Chinese family business, *International Journal of Psychology*, 25, 593-618.

Ingram, L. C. (1980). Leadership and democracy: Further observations on the pastoral role in congregational churches, an association paper of Southwestern Sociological Association.

Jaggi, B. (1977). Job satisfaction and leadership style in developing countries: The case of India, *International Journal of Contemporary Sociology*, 14(3-4), 230-236.

Jurma, W. E. (1979). Leadership structuring style, task ambiguity, and group member satisfaction, *Small Group Behavior*, 9(1), 124-134.

Kalleberg, A. L. (1977). Work values and job rewards: A theory of job satisfaction, *American Sociological Review*, 42, 124-143.

Kalleberg, A. L. (1987). Organizational commitment in Japan and America, *Social Science*, 72(2-4), 159-162.

Kalleberg, A. L. (1990). The comparative study of business organizations and their employees: Conceptual and methodological issues. In C. Calhoun (Ed.), *Comparative social research: Business institutions* (pp. 153-176). Greenwich, CT: JAI Press.

Lee, S. T., and Tu, C. C. (1991). The concept of organizational commitment and its measurement, *Chiao Ta Management Review*, 11(1), 51-74.

Li, J., Koh, W. K., and Hia, H. S. (1997). The effects of interactive leadership on human resource management in Singapore's banking industry, *International Journal of Human Resource Management*, 8(5), 710-719.

Lincoln, J. R. (1989). Japanese organizations and organizational theory. In B. M. Straw and L. L. Cummings (Eds.), *Research in organizational behavior*, vol. 12. Greenwich, CT: JAI Press.

Lincoln, J. R., and Kalleberg, A. L. (1985). Work organization and workforce commitment: A study of plants and employees in the United States and Japan, *American Sociological Review*, 50, 738-760.

Marsh, M. K., and Atherton, R. M., Jr. (1981). Leadership, organizational type, and subordinate satisfaction in the US Army: The Hi-Hi paradigm sustained, *Humboldt Journal of Social Relations*, 9(1), 121-143.

McElroy, J. C., and Shrader, C. B. (1986). Attribution theories of leadership and network analysis, *Journal of Management*, 12,(3), 351-362.

Minturn, L. (1995). Communes as moralnetts, *Cross-cultural Research*, 29(1), 5-13.

Mortimer, J. T., and Lorence, J. (1979). Work experience and occupational value with socialization: A longitudinal study, *American Journal of Sociology*, 84, 1361-1385.

Mowday, J. T., Porter, L. W., and Steers, R. M. (Eds.) (1982). *Employee-organization linkages: The psychology of commitment, absenteeism and turnover*. New York: Academic Press.

Orpen, C., and Hall, C. (1994). Testing the substitute model of managerial leadership: An examination of moderators of the relationships between leader reward and punishment behaviors and employee job satisfaction, *Studia Psychologica*, 36(1), 65-68.

Porter, L. W., and Steers, R. M. (1973). Organizational, work, and personal factors in employee turnover and absenteeism, *Psychological Bulletin*, 80, 151-176.

Porter, L. W., Steers, R., Mowday, R., and Boulian, P. (1974). Organizational commitment, job satisfaction, and turnover among psychiatric technicians, *Journal of Applied Psychology*, 59, 603-609.

Portes, A. (1998). Social capital: Its origins and applications in modern sociology, *Annual Review of Sociology*, 241-24.

Podsakoff, P. M., Niehoff, B. P., MacKenzie, S. B., and Williams, M. L. (1993). Do substitutes for leadership really substitute for leadership? An empirical examination of Kerr and

Jermier's situational leadership model, *Organizational Behavior and Human Decision Processes*, 54, 1-44.

Rao, A., Thornberry, N., and Weintraub, J. (1987). An empirical study of autonomous work groups: Relationships between worker reactions and effectiveness, *Behavioral Science*, 32(1), 66-76.

Schnake, M., Cochran, D., and Dumber, M. (1995). Encouraging organizational citizenship: The effects of job satisfaction, perceived equity and leadership, *Journal of Managerial Issues*, 7(2), 209-221.

Schwartz, P., and Ogilvy, J. (1979). *The emergent paradigm: Changing patterns of thought and belief*. Menlo Park, CA: SRI International.

Shamir, B. (1986). Meaning, self and motivation in organizations, *Organization Studies*, 12(3), 405-424.

Simagin, Y. A. (1996). Managerial style and the effectiveness of scientific groups, *Sotsiologicheskie Issledovaniya*, 23(3), 129-133.

Stogdill, R. (1974). *Handbook of leadership*. New York: Free Press.

Yoon, J., Backer, M. R., and Ko, J. W. (1994). Interpersonal attachment and organizational commitment: Subgroup hypothesis revisited, *Human Relations*, 47(3), 329-351.

Young, R. S., Chen, W. C., Lee, S. T., Young, A., Ho, C. M., Tu, C. C., and Ohn, C. S. (1991). *A study of employees' willingness to work and behavioral motivation* (in Chinese), a bulletin report of the Sun Yat-sen Center of Social Policy Studies, National Sun Yat-sen University, Kaoshiung, Taiwan.

Appendix

Table 6.A1 Gamma score indicating the relationship between leader ability and work problem

	Work overload	*Work inequity*	*Role ambiguity*
Overall abilities	-.243**	-.141**	-.199**
Technical	-.239**	-.177**	-.240**
Organizational	-.180**	-.111**	-.147**
Scheming	-.192**	-.146**	-.175**
Motivating	-.175**	-.124**	-.149**
Self expression	-.198**	-.159**	-.213**
Human relation	-.222**	-.184**	-.218**
Rewarding	-.217**	-.149**	-.179**
Problem solving	-.193**	-.169**	-.211**

*p<.05 **p<.01

Table 6.A2 Gamma score indicating the relationship between leader attitude and work problem

	Work overload	*Work inequity*	*Role ambiguity*
Overall attitudes	-.234**	-.140**	-.203**
Understanding	-.278**	-.228**	-.285**
Fair -.183**	-.152**	-.163**	
Justice	-.182**	-.123**	-.159**
Consideration	-.209**	-.121**	-.166**
Patience	-.184**	-.102*	-.164**
Communication	-.092**	-.097	-.092

*p<.05 **p<.01

Table 6.A3 Gamma score indicating the relationship between leader ability and work problem in production divisions

	Work overload	*Work inequity*	*Role ambiguity*
Overall ability	-.282**	-.119**	-.214**
Technical	-.283**	-.172**	-.278**
Organizational	-.230**	-.107*	-.183**
Scheming	-.226**	-.104*	-.181**
Motivating	-.202**	-.096*	-.110
Self expression	-.219**	-.119**	-.214**
Human relation	-.235**	-.190**	-.234**
Rewarding	-.234**	-.139**	-.216**
Problem solving	-.206**	-.174**	-.236**

*p<.05 **p<.01

Table 6.A4 Gamma score indicating the relationship between leader ability and work problem among workers

	Work overload	*Work inequity*	*Role ambiguity*
Overall abilities	-.283**	-.122**	-.217**
Technical	-.292**	-.160**	-.269**
Organizational	-.235**	-.121**	-.171**
Scheming	-.227**	-.169**	-.205**
Motivating	-.202**	-.108*	-.164**
Self expression	-.211**	-.156**	-.231**
Human relation	-.223**	-.150**	-.189**
Rewarding	-.247**	-.152**	-.216**
Problem solving	-.222**	-.133**	-.222**

*p<.05 **p<.01

Table 6.A5 Gamma score indicating the relationship between leader attitude and work problem in production divisions

	Work overload	*Work inequity*	*Role ambiguity*
Overall attitudes	-.238**	-.129**	-.214**
Understanding	-.270**	-.162**	-.245**
Fair -.193**	-.149**	-.186**	
Justice	-.175**	-.115*	-.161**
Consideration	-.243**	-.130**	-.194**
Patience	-.206**	-.114*	-.196**
Communication	-.058	-.077	-.111

*p<.05 **p<.01

Table 6.A6 Gamma score indicating the relationship between leader attitude and work problem among workers

	Work overload	*Work inequity*	*Role ambiguity*
Overall attitudes	-.250**	-.130**	-.209**
Understanding	-.255**	-.195**	-.251**
Fair -.216**	-.167**	-.200**	
Justice	-.212**	-.130**	-.174**
Consideration	-.258**	-.133**	-.206**
Patience	-.204**	-.121*	-.194**
Communication	-.095	-.038	-.048

*p<.05 **p<.01

Plate 8 Clayoquot Sound ➧

7

The Globalization of Political Space: Reflections on Clayoquot Sound

Warren Magnusson

Professor, Department of Political Science, University of Victoria

Many people have remarked on "globalization," the recently fashionable term for describing the processes that seem to have thrust us all into a new space. This is a space where people in the world are present to one another, involved with each other, dependent on one another in ways that were scarcely imaginable even when the Americans were already on the moon. Obviously, the development of electronic media of communications has had a profound effect: you and I can sit at our computer screens or in front of our television sets or simply pick up an old-fashioned/new-fashioned telephone and be in touch instantaneously with people and events in other parts of the world. However, it is the use of these media by modern businesses that is most striking. Even small-scale businesses can operate on a global scale, and organize production, distribution, and finance in a way that maximizes locational advantages. As a result, it has become increasingly difficult for governments to control businesses. The demands of globalized businesses weigh ever more heavily on the minds of politicians, not only because the politicians themselves are afraid of capital flight, but also because ordinary people have been sensitized to the thought that the favours of capital are as easily withdrawn as conferred. At the same time, repressive governments, like the one in mainland China, find that they are faced with popular resistance that is difficult to monitor and even more difficult to control, because it is facilitated by the globalized media of communications. As a result, there have been rumblings from the authorities everywhere about the "death of the nation-state" and the emergence of a "new world order."

It is not my purpose to offer another account of globalization—except to say, in passing, that recent developments are just the latest phase in a process that has been going on for at least five centuries (and arguably for thousands of years before that). Instead, I want to raise some issues about the political space we Vancouver Islanders confront. I do this through some reflections on the politics of Clayoquot Sound, a relatively remote place on this island of mine (just south of the site where a number of undocumented boat people from South China landed in the summer of 1999). Clayoquot had its 15 minutes of world fame in the summer of 1993, when (according to our national news magazine) the "whole world" was watching the ongoing protests there against imminent clearcut logging. I doubt if many people in Taiwan heard about the protests—or could care less about them, if they did—but they were a big event on Vancouver Island: probably the biggest political event of the year. The protests were significant not only because they involved a remarkably sustained (and highly photogenic) campaign of civil disobedience, but also because they spilled beyond the island, the province, and the country as a whole. The provincial government thought that it had the issue under control, when it made a decision in April that year about which parts of the area were to be opened for logging, which parts were to be preserved as wilderness, and which parts were to have "special treatment" (that is, somehow being logged and not logged at the same time). However, the government's decision not only prompted protests on the ground, in Clayoquot, but also led to an international campaign that the provincial government could not control. The province was drawn into a political space where a party's image in the international media was of decisive importance, and where market reactions were more significant than responses within the domain of the state system. In that political space, the government found that its ostensible sovereignty with respect to the land in question was only one political asset among many, and not really the crucial one.

For me, the frustrations of the provincial government of British Columbia with respect to the issues at Clayoquot are like a parable of contemporary politics—a parable that is as relevant for Taiwanese as it is for Canadians. It is beyond my competence to make any systematic comparison between Taiwan and Vancouver Island, but I do want to suggest that contemporary politics has certain features that are recognizable everywhere, and that the categories we have been utilizing to comprehend these features are profoundly inadequate. I cannot develop my argument fully in this chapter, nor can I give an adequate account of the example to which I will be referring. Instead, I propose to focus on three issues related to the globalization of political space, issues that may all be illustrated by struggles in and around Clayoquot Sound. These are: (1) the articulation of "the market" as a political space; (2) the proliferation of "identities" in political struggle; and (3) the transfiguration of sovereignty as a disciplinary discourse. Briefly, I will be arguing that the market has emerged as a decisively important political space, and that it has done so despite—I am tempted to say "because of"—the efforts of neo-liberals to de-politicize it. Ironically, efforts to "tame the state" tend to displace politics into the space of the market, and so politicize what free marketeers had hoped would be a protected space governed by apolitical rules of economic rationality. I will also be arguing that the disruption of the nation-state as the privileged site for politics is both cause and effect of the proliferation of political identities—identities that are by no means "natural," but are rather to be understood as effects of political struggle. Finally, I will suggest that the apparent displacement of the state as a site of ultimate political resolution is by no means a sign the discourse of sovereignty itself has been displaced. On the contrary, I will argue, sovereignty discourse remains predominant, despite the transfiguration implicit in de-centring the state.

The Market as a Political Space

Let us begin with the market, since it is the shift from state regulation to market exchange that has caught the attention of most analysts. Economists have long recognized that the market is in principle global. In their view, states and societies exist within a natural order given by the logic of exchange. Although peoples and governments can defy that order, by setting up barriers between themselves and the outside world, the price of such defiance is high. To receive the full benefits of trade (so the economists say), a country must open itself up, and participate fully in the exchange economy. This means becoming specialized in particular forms of production and learning to service faraway markets. The rules of the game are the ones implicit in the logic of exchange. The laws must secure contracts, protect private property, allow for entrepreneurial freedom, and so on. Otherwise, the country concerned will deny itself the full benefits of the exchange economy, and invite reprisals from other countries that play by the rules. The implication of this is that there is relatively little scope for political, social, and cultural differentiation within a rationally organized global economy. We are all free to choose, but we all have to choose the same things.

Liberals have resisted this conclusion. They have pointed to two positive features of the exchange economy. The first is that the rules that protect private property, secure contracts, and allow for economic enterprise themselves provide for important freedoms—freedoms that may facilitate the expression of political, social, and cultural differences. In an exchange economy, an entrepreneur can start a Cantonese newspaper or a television station in Vancouver, and in so doing cater to the needs of Cantonese-speakers living in Canada. This may help to preserve cultural differences that would otherwise be lost. Similarly, entrepreneurs may establish shopping centres, housing complexes, business facilities, schools, medical services, and other facilities that cater to a Cantonese-speaking population in a country where English is otherwise the predominant language. A government dominated by English-speaking people is unlikely to be so accommodating. So, in this respect as in others, the free market may secure the conditions under which cultural, social, and political differences can be sustained. (Political organizations that promote the interests of Cantonese-speakers are simply another form of enterprise secured by market freedoms.)

The second positive feature of the exchange economy is implicit in the first: namely, that the effort to appeal to particular groups of consumers entails an attempt to identify ever-more-minutely distinguished groups, whose needs can be met by specialist producers. Thus, market-oriented enterprise tends to produce goods and services that satisfy the requirements of consumers who differentiate themselves from one another in terms of their culture, tastes, and lifestyle. This enables both an individualized "mix-and-match" (for instance, a person may choose to practise Zen Buddhism, adopt an Indian vegetarian diet, live in a California-style house, immerse herself in Irish writing, and make a living selling handicrafts collected in Africa and South America) and a collectivized commitment to particular forms of community (for instance, the Hassidic communities in many large cities or the even more ubiquitous gay communities set themselves apart in various ways, without relying on state power). Thus, the logic of free market exchange seems to encourage ever more varied expressions of cultural difference.

It would be nice to think that the processes we have been experiencing are completely benign. However, the forms of difference enabled by the market are quite tightly constrained. To see that this is so, we need only ask a simple question: "Is there greater cultural diversity in the world now, in 1999, than there was in 1499? Or, is there less difference?" It seems obvious to me that the answer is, "There is less." And, if we ask why, we are soon led to the "rules of the game" that the economists have tried to describe for us. To participate fully in market exchange, people must shed their old ways of life, and learn to orient themselves toward the global market. This means becoming competitive individuals, and abandoning beliefs that tie a person to old ways, old places, and old communities. In becoming other than what our ancestors were, we become more and more like people from other cultures who live under similar conditions. So, the difference between an Italian and an Englishman or between a Korean and an Indian becomes more and more superficial. The differences supported by a globalized exchange economy are more like the ones celebrated in Canadian multi-culturalism (that is, we have our national dishes, special costumes, and traditional songs, but under the skin we are all much the same) than the deep differences that the Europeans encountered when they first began to trade on a global scale five centuries ago.

To note this is not to suggest that we can go back to the days of geographically isolated cultures—or even that we would want to do so. I simply want to underscore the fact that the homogenizing effects of the exchange economy are much more pronounced than the ones that encourage diversity. The diversity facilitated by an exchange economy is real enough, but it is limited by the very logic of exchange relations, relations that require people to *be*—and hence to *think*—in particular ways. One cannot be both a traditional Nuu-chah-nulth person *and* a successful participant in the global economy. The latter identity entails abandonment of all but the bare cultural remnants of the former. So, it is a delusion to think that the freedoms implicit in a globalized exchange economy are ones that enhance cultural difference. Everything economists say about the way a rationally organized market has to work points toward a requirement for cultural, social, and political uniformity. All the evidence of the past five centuries suggests that greater uniformity is indeed the main effect of the globalization of exchange relations.

Friedrich von Hayek has offered the most complete account of the logic of exchange relations. He and his followers have argued forcefully against efforts to impose a different sort of order on the world, an order corresponding to some vision of the way the world ought to be. In his view, such efforts always produce more harm than good. A preconceived order is only as capacious as the imaginations of the people who make it—and that is not very capacious, since we all have quite limited visions. Any order that we impose on the world will stifle initiative, block innovation, and impose uniformity. It can only be maintained by violence. It will be rife with corruption, and it will breed more violence. By contrast, the unplanned order generated by market exchange will be an effect of the free activity of millions of individuals with different ideas and different ambitions. What emerges will be a dynamic, self-regulating, self-correcting system that meets people's needs more effectively than anything that could be conceived and implemented as a preconceived plan.

A Hayekian order is supposed to be free from politics, in the sense that it secures the constitution of liberty naturally, as an unplanned outcome of human cooperation. Perhaps only an economist could take such a preposterous theory seriously—but perhaps not. Hayekian ideas are now quite popular among those who have taken an interest in self-organizing systems and spontaneous order: phenomena that have been identified at the molecular level, in weather systems, in the cells of living organisms, and so on. In a way, the idea of a self-organizing system is quite appealing, because it absolves us of responsibility for the arrangements under which we live. Ironically, Hayek's own writing is filled with moral and political imperatives: that is, orders to the participants in self-organizing systems about what they should and should not do. Similar orders are now being issued every day, by the bond-rating agencies, investment houses, and market analysts, all of whom have very precise ideas about how we should behave, what we should expect of our governments, and, indeed, how we should live our day-to-day lives. Nowhere is there such insistent moralizing—and such consistent political propaganda—as in the business pages of the daily newspaper, where we are told day after day that governments have to cut taxes, de-regulate business, privatize public services, reduce trade barriers, and so on. The moral indignation of the average business writer faced with a high tax rate is much greater than that of a Sunday preacher faced with the usual litany of human sins.

When people rise up in moral indignation to condemn behaviour that they think is at odds with the natural order, we have good reason to suspect that what they describe as unnatural is in fact all too common. Vegetarians may like to think that it is unnatural for people to eat animals, but it is quite obvious that people have been eating animals for thousands of years. It seems that carnivorousness is quite natural to us humans. So too, is resistance to the logic of the market. The unusual thing is not that people today should want to live in ways that are at odds with freedom of exchange. What is truly unusual is that people should accept that their lives should be governed by the logic of exchange. The indignation of the business press is one sign that people are actually quite resistant to the idea that their fate should be determined by the vagaries of consumer demand and technological innovation. Ordinary people tend to reject the idea that human welfare can be measured in strictly economic terms, and they are attached to the notion that there are communities, cultures, religions, and ways of life that ought to be protected *again*st the logic of free exchange. This makes business commentators *very* angry, because they think that such ideas are completely irrational.

I will say more below about the way that the exchange economy has been constructed, politically. Suffice to say here that an economy as complicated as ours only can work if there are precise rules about what can be owned, what can be bought and sold, what constitutes a fair exchange, what counts as theft, and so on. However natural the trucking and trading of Adam Smith's men may appear, there is nothing very natural about Microsoft's ownership of the codes that enable my computer to operate. It takes a great leap of the imagination to say that a person can own a computer code or a particular squiggle that denotes a familiar brand. It takes a similar leap of the imagination to suppose that Microsoft is an artificial person with most of the rights that you and I have. To suppose that Microsoft's property rights are simply the effects of a self-organizing system at work is to ignore the *political* activity that went into persuading judges and legislators to recognize limited liability companies as artificial persons and to treat ideas and symbols as things that could be bought and sold. It might have been otherwise. The authorities might have decided that there could be no such thing as an artificial person or intellectual property. There would have been nothing unnatural about such a result. However, things went the other way, for reasons that we can explain in terms of the interests who were engaged in business and controlled large amounts of property. To suppose that what resulted was simply for the best, in this best of all possible worlds, is to revert to a Panglossian optimism that we could well do without.

A much more realistic account of exchange relations is one that recognizes, first, that all exchange relations are socially embedded, and, second, that all social relations are politically inflected. Hayek's contemporary, Karl Polanyi, is perhaps most famous for making the argument that the *laissez-faire* economy had to be *created*: that it was by no means natural to people, that it had to be formed by imposing things on people against their will, and that this imposition required the use of state power. Polanyi is also famous for his insistence that markets only work in a socially beneficial way when they are embedded in a wider set of social relations that sustain the morality and the attitudes necessary for the smooth conduct of market activity. As contemporary Polanyians point out, the social relations necessary for a successful market economy seem to be absent in places like Russia, where there is too little mutual trust and too little social solidarity for people to accept market gains and losses as natural. If social relations are characterized by a deep mistrust, a good bargain will appear like theft, and many people will be tempted to use fraud, extortion, and bribery to make gains. Such behaviour is perfectly natural, but it is not what Hayek and his followers have in mind when they praise self-organizing systems. Sierra Leone and Angola are not usually mentioned when people sing the praises of open competition. What the free marketeers have in mind is in fact a *highly regulated* system of exchange, secured as necessary by force of arms. As any competent political analyst knows, sheer force is always insufficient. The desired system works best when ordinary people have internalized the norms associated with it. Hence, the need for daily propaganda, to remind us of how we must behave in the global economy.

If the market only works as expected in a particular social context, and if that context has to be produced and sustained *politically*, then it is clear that the market is both an object of political action and a space within which politics occurs. People actually find it easier to accept the first conclusion than the second. That is because—despite all the talk about self-organizing systems—people still tend to think that everything will fall apart if there is no "law and order," and they suppose that governments with a monopoly of the means of extreme violence are necessary to provide law and order. (Such a view is actually implicit in Hayekian theory.) If governments are necessary, states are necessary, and if states are necessary there will necessarily be politics. Only the most fanatical believe that there is just one way of organizing an exchange economy. Indeed, the general theory that Hayek expounds would seem to imply that people in different parts of the world should be free to experiment with different arrangements. Presumably, the most successful arrangements would then win out, and be copied elsewhere. In any particular polity, there would presumably be different ideas about the ground rules of economic exchange, and people would be free to expound their ideas in competition with one another. Thus, the state would be constituted as a political space, in which different visions of the free society would be expounded.

It is a relatively easy step to say that interstate relations are an extension of the political space of the state. Insofar as the rules that govern the global economy are worked out by negotiations between states, or articulated by trans-national bodies established by treaty, then it is easy enough to see that there is politics associated with these processes. There is an incipient trans-national political debate between those, like George Soros, who believe that the global economy has to be more tightly regulated by trans-national institutions of government, and those who (following Hayek) express continued suspicion about the idea of governmental "interference" in the economy. One can imagine a 21st century politics that replicates internationally some of the features of the left-right struggle that has characterized 20th century domestic politics. This makes some people uneasy, especially those who have applauded the success of contemporary business in ridding itself of many of the strictures imposed by national governments in the first decades after the Second World War. However, there is another dimension of the political struggle that makes proponents of the "free market" even more uneasy: namely, the politicization of exchange relations.

One of the premises of Hayekian theory, and indeed of liberal theory in general, is that "the economy" is to be understood as something separate from "the state" and hence from politics. Exchange relations are to be governed by economic reason, not by political theory. That the state might regulate the economy in certain ways is perhaps tolerable. That the state should secure the economy is necessary. But, that the line between politics and economics or between the state and the market should be erased is, from a liberal perspective, highly dangerous. Thus, the idea that a shareholder's meeting or a meeting of a company's board of directors should be a forum for political debate, in which different ideas about the future of the world are articulated and a policy is decided, is anathema. The politicization of such meetings erases the distinction between public and private affairs, and introduces a much wider range of considerations into economic decision-making. If economic decisions are to be considered as expressions of our values, and if our values are to be considered proper subjects of political deliberation, then the line that separates economics from politics begins to seem artificial. Some human relations may be mediated by the market, but there are many other situations in which people settle their differences politically. Politics is often characterized by argumentation and deliberation, but there is much more to it than that. People appeal to various loyalties, raise hopes, nurture grievances, articulate visions, mobilize support, threaten reprisals, use agreed procedures for dispute settlement, and (in a pinch) impose decisions on one another forcefully. All of this can be characterized as self-organizing activity of a kind that is perfectly natural for humans, but does not conform to anyone's ideal of how we should behave—least of all, Hayek's.

Hayekians might like people to refrain from politicizing the market, but (ironically) Hayekian policies are forcing people to do exactly that. The recent trend has been to force states to abide by a set of rules that constrain their capacity to interfere with business. The WTO, or some similar body, now rules out policies that might have been implemented by a national government 20 years ago. Even policies that are technically allowable are not politically feasible, because international investors punish governments that deviate from the norms articulated by the international investment community. (Marx would have enjoyed describing all of this.) In the circumstances, the advocates of other views have little option but to claim the market as a space in which they can legitimately operate. Much attention has been given to the way in which critics were able to mobilize against the Multilateral Agreement on Investment. Some credit defeat of the MAI to successful trans-national political organizing. Be that as it may, such organizing is probably less significant than another form—one that takes the struggle right into the market.

Let me refer to my Clayoquot example. Opponents of logging in the region used various tactics, but their decisive move was to launch what they called their "markets campaign." The idea was simple. The provincial government, which might have intervened to prevent logging, was reluctant to do so because it would have had to pay compensation to the company that held the timber rights, and it would have had to provide alternative employment for workers who lost their jobs as a result of the closure of logging operations. Neither measure seemed feasible to the government, which was in any case afraid of setting expensive precedents for other parts of the province. There was also significant political opposition to a logging ban from people who made a good living from the industry, and who feared that other economic alternatives would not be so beneficial for them. So, to get the logging stopped, environmentalists had to appeal over the heads of the provincial government *and* over the heads of the public in British Columbia to a broader public, less concerned about the local costs of environmental preservation. This public was constituted not as an electorate, but rather as a *consumerate*. To stop the logging in Clayoquot, environmentalists had to persuade the people who purchased wood products to seek supplies from other, more acceptable sources. As the wood was a raw material used in the production of various end-products (chiefly paper), the immediate purchasers were profit-seeking companies and not ordinary consumers. The companies could be expected to follow the logic of lowest-cost sourcing for their inputs, *unless* the end-consumers of their products had serious objections to the companies' purchasing practices. To succeed in their markets' campaign, the environmentalists had to generate apprehension among the relevant *corporate authorities* about the possible effects of a consumer boycott of products made from the Clayoquot woods.

There were a number of dimensions to the markets' campaign. Firstly, publicists for the anti-logging cause toured Europe, and held public meetings to air their views about the issues. These tours could be conducted at relatively low cost because there was an existing network of environmental organizations ready to receive the campaigners, organize meetings, supply accommodation, and transport people from city to city. Secondly, the campaigners attracted the attention of sympathetic television producers and documentary filmmakers who were prepared to make critical programs about logging practices in British Columbia in general and Clayoquot Sound in particular. These programs could attract a fairly large audience, because the site at stake (Clayoquot Sound) is exceptionally photogenic, and the issues (of wilderness preservation and aboriginal rights) resonate well with an urban audience. Thirdly, the Clayoquot campaigners were able to persuade major international environmental organizations to highlight the situation at Clayoquot in the context of a broader campaign to save the world's remaining rain forests. As a result, these organizations bought advertising space in major publications like *The New York Times*. The logging companies were forced to respond, and *that* made the issue "news." The consequence was a spate of stories that provided free publicity for the campaign. Fourthly, the campaigners used their own civil disobedience—which resulted in more than 900 arrests over a 3 month period—to attract ongoing television news coverage and to publicize their own determination. The publicity continued during the trials of those arrested. In related moves, the campaigners took advantage of their support among celebrities to draw attention to the dispute, and used the Peace Camp at Clayoquot as a site for a free rock concert, broadcast on MTV and MuchMusic. Finally (and in the context of these other moves), the campaigners began asking companies to stop purchasing paper or other products made from wood extracted from the area. They signalled to the logging companies and to the British Columbia government that demands for a wider boycott of BC forest products would be forthcoming if there was no satisfactory response to the issues at Clayoquot.

The markets campaign had an immense impact. Whether the Clayoquot protesters would have been successful in organizing an international boycott of Clayoquot forest products is doubtful. However, there were a number of major companies—including *The New York Times*—that chose *not* to test the environmentalists' resolve. In highly competitive markets, relatively small shifts in consumer preferences can lead to major profit losses. Why risk public disapproval for buying "tainted" products, when there are alternative sources of supply? Even if only a few major purchasers decided to "greenwash" themselves by adopting the environmentally correct stance, the opponents of logging could use this fact to spook the logging companies and the government that supported those companies. Clearly, profit margins and public credibility were both at risk for companies that had to fight protesters to get the logs that they wanted, and fight them again to get their products into the markets. It was not impossible for internationally networked environmental organizations to create adverse publicity at every major point in the chain of production and distribution. Giving up on Clayoquot was a relatively small price to pay for peace—and that is the price that the major company involved eventually agreed to pay. (The company then linked this change of heart to other changes in its logging practices, and presented itself to the public as the "greenest" of the province's lumber companies.)

I do not mean to suggest that the environmentalists' campaign at Clayoquot was an unadulterated success for them. The logging companies were fairly successful in their efforts to deflect and contain the pressure generated. However, it is not an assessment of who won and who lost that concerns me here. My point is simply that the market itself emerged as the terrain of political struggle: the place where battles were won and lost. The hearts and minds of consumers was the ultimate prize. Businesses, with their immense resources and long experience with advertising, had considerable advantages in any political struggle that arose within the market. On the other hand, these same businesses were sensitive to small shifts in consumer sentiment. Environmental organizations had enough resources to mobilize their own constituency, and this mobilized constituency posed a political threat, not so much because of the force it could generate or the elections it could win, but because of the money it had to spend on the products the businesses were trying to sell. Green marketing is often deceptive, but the fact that it occurs is an important sign: a sign that the market itself has been recognized, on both sides, as a field in which political struggles will inevitably occur.

There are many more famous examples than Clayoquot of the politicization of market relations: the decades-long boycott of South African products and the related campaign for disinvestment; the effort to organize a grape boycott in support of agricultural workers in California; boycotts of Shell and Nestle, in response to their activities in West Africa; attempts to develop alternative markets for agricultural producers (like coffee-growers) in Latin America; purchases of wildlife reserves by the Nature Conservancy and other environmental organizations; global campaigns against nuclear energy, CFC production, whale hunting, and so on. The Clayoquot case is just another example of what happens when people find "normal" political channels ineffective. If my overall argument is correct, one effect of the shift to the market will be a corresponding shift in the locus of political activity. More and more disputes will be played out as struggles for the hearts and minds of consumers. No doubt, consumers will often react the way business writers want them to. On the other hand, there is little indication that people are actually prepared to give up being human beings, and human beings still have concerns that are at odds with the logic of market exchange.

The Proliferation of Identities

One interesting feature of the struggles in and around Clayoquot Sound was that the authorities—both corporate and governmental—were so desperate to contain the proliferation of political identities. If people had just kept to the identities that had been previously established, things would have been much easier to manage. However, the dynamics of the situation were such that identity confusion was more the rule than the exception. Let us explore the facts more carefully.

The division between market and state produces two identity sets. On one hand is a set of economic identities: worker, consumer, investor, employer, *etc.* On the other hand is a set of political identities: citizen, voter, representative, governor, *etc.* According to the ideology of liberalism, the primary identity from which all others arise is "the individual." Economy and polity are to be organized in a way that maximizes the wellbeing of individuals. The differentiation of individuals from one another—hence the proliferation of identities—can be accommodated to a considerable degree without disrupting the state or the market. Nonetheless, it is easier to manage people if they conform to standardized identity models: that is, if they relate their various identities to one another in predictable orders of priority. For instance, a "family man," who understands that his first duty is to provide for his wife and children, may be a particularly reliable employee who will be loathe to take risks that might harm his good relations with his employer. If such a person also understands that his main duty as a citizen is to obey the law, there may be little tension between the requirements implicit in the "family man's" various identities. "Family man" is thus (in this instance) an architectonic identity that resolves various tensions, and enables the person concerned to function smoothly within both the economy and the polity. Smooth-functioning people, with predictable architectonic identities, are relatively easy to govern, and easy to organize economically.

As many commentators have noted, liberalism tends to dissolve intermediary identities. As it is the individual who has ultimate value, and social arrangements are simply means to individual wellbeing, there is little reason for any particular individual to sacrifice his or her welfare for the benefit of some intermediary group (be it family, community, company, or nation). The strongest claims on the individual are those associated with the authority that makes individual freedom possible, be it the state, the market, or some combination thereof. So, the liberal order tends to make duties general and abstract: play by the rules, obey the law, and so on. According to liberal theory, particular individuals are free to be what they want to be: that is, to organize their identities according to their own preferences. What follows, logically, is that there should be a play of identities within and among individuals. If a person is to discover his or her identity, s/he must try out a number of possibilities and test them against one another. Moreover, s/he must feel free to reject the standardized identities that make government easy and allow for the efficient organization of production. To put it simply: if people were to behave the way liberal theory suggests, neither the state nor the market would be orderly.

During the struggles at Clayoquot Sound, the authorities relied on some familiar hierarchies of identity. With respect to political loyalties, people were expected to accept the idea that they were Canadian citizens first, British Columbians second, members of local communities third, and adherents of particular parties, movements, or interest groups last. If people understood themselves in these terms, they would accept that their provincial government had the right to decide what would happen to the land under its jurisdiction and that the courts and the police had the responsibility to enforce that decision. With respect to personal interests, the authorities expected people to accept the idea that their most important interests were economic, and that economic interests could be understood in terms of opportunities for employment and investment. In other words, people were to see themselves in terms of the range of identities that economists have described, and accord those identities an importance comparable to the ones implicit in the hierarchy of political loyalties. There might be some tension between economic interests and political loyalties. However, a good government would resolve that tension by acting in the economic interests of its constituents. This is what the Government of British Columbia attempted to do.

The interesting feature of the struggles at Clayoquot is that they brought out so many contrary identities, identities that could not be so easily managed. Let me review the most important of these identities.

(1) Aboriginality

The Nuu-chah-nulth people claimed the entire territory belonged to them, and that it had belonged to them since time immemorial. Consider the following:

Declaration

Let it be known as of April 21, 1984, we the Clayoquot Band, do declare Meares Island a **Tribal Park**.

1) Total preservation of Meares Island based on title and survival of our Native way of life.
2) Preserve Meares Island as an economic base of our people to harvest natural, unspoiled Native foods—including all:
   a. seafoods and shellfish
   b. protect our traditional hunting rights of deer and waterfowl, etc.
   c. protect the right of our elders to continue gathering their Indian medicines.
   d. protect the right of Native artists to continue the gathering of their needs. Cedar bark, cedar for canoes, paddles, and masks, etc.
3) Protection of all salmon streams on the Island.
4) Protection of all herring spawning areas around the Island.
5) Protection of all traplines.
6) Protection of all sacred burial sites on Meares.

The Native people are prepared to share Meares Island with non-Natives, providing that you adhere to the Laws of our Forefathers; which were always there. On this basis, we recognize your needs for:

1) a. Watershed, as they already have in place their water system on Meares Island.
   b. Hunting of waterfowl in Lemmens Inlet.
   c. Existing mariculture leases. We would reserve the right to process any further development, be it watershed or further maricultural leases.
2) We would permit access to the Island for recreational purposes—hiking, camping, fishing, whale watching, gathering restricted amounts of seafoods and shellfish.
3) Recognize our Land Claims and that there be no resources removed from Meares Island excluding watershed.

Signed by: George Frank (Hereditary Chief), Alex Frank Sr. (Hereditary Chief), and the Clayoquot Band Council

The Clayoquot are one of the 14 tribes who describe themselves collectively as the Nuu-chah-nulth people. The statement above was issued 9 years before the blockade that brought so much attention to the region in 1993. In 1985, the Clayoquot obtained a court injunction that prevented logging at Meares Island (in the middle of Clayoquot Sound) pending resolution of the dispute between them and the Government of Canada with respect to aboriginal rights in the area. Throughout the subsequent disputes in the region, the provincial government and the forest companies were well aware that the Clayoquot (or other Nuu-chah-nulth tribes) might seek to enlarge the scope of the injunction.

For present purposes, the interesting feature of this declaration is what it tells us about identities. The identities asserted here are not ones that conform to the requirements of the Canadian state *or* the global market.

(2) Femininity

Again let me quote, this time from a remarkable memorandum written by Nancy Scott, an employee of the major logging company, MacMillan-Bloedel. In it, she muses about the problems that the company faces in dealing with what she perceives as a shift from "masculine/patriarchal" to "feminine/matriarchal" values. As she says:

> *We stand accused of 'raping and pillaging.' Our activities are described as 'skinning the earth alive' (a masculine hunting metaphor) resulting in an 'environmental holocaust' (war is a classic masculine metaphor).*
>
> *Clearcutting and slashburning are viewed as acts of war against helpless Mother Earth. As an industry we talk with pride of our feller-bunchers and grapple yarders. The environmental movement talk of 'Gaia,' a semi-mystical term for the planet.*

She sums up the competing values as follows:

Patriarchal/Masculine	*Matriarchal/Feminine*
Aggression	Consensus
Domination	Nurturing
Achievement	Sustainable
Development	Preservation
Growth	Conservation
Action	Choice
Results	Process
Data	Emotion
Representation	Participation
Profit	Sharing
Progress	Status Quo
Hierarchy	Network

Then she applies the same dichotomy to logging practices:

Clearcutting	Selective logging (sharing)
Logging (domination)	Planting (nurturing)
Slashburning (fire)	Farming
Working Forest	Wilderness
Reserve	Sanctuary
Waste	Salvage (conservation)
Log Exports	Value-added (keeping it in the family)
Multinationals	Share groups
Corporate decisions	Public involvement (consensus)

Scott goes on to suggest that the company respond to the challenge by re-describing its activities ("Let's characterize ourselves as parents caring for sick trees instead of soldiers going to war against pests. Perhaps we can be doctors of the forests instead of its [*sic*] managers"), putting women out front as spokes-people (a little self-interest here?), and changing some logging practices (for instance, by promoting the recycling and salvage of wood waste). That she sees this mainly as a public relations exercise is obvious, but again one is struck by the fact that she is dealing with a challenge to the very identities that under-pin the state and the market.

The play of identity at Clayoquot was exceptionally complicated. Union-ized loggers, who had often been at odds with their employers, came out in strong support of the company's right to harvest timber. Their wives appeared at the environmentalist blockade to chastise the protesters for threatening their homes and families. Anyone from outside the region (and especially from outside the country) was condemned as a meddler. But then the environmen-talists presented themselves as defenders of a land that belonged at one level to the aboriginal people and at another level to all the beings that inhabited the area. From this perspective, the loggers were the intruders. The environmen-talists also linked up with local people who had an interest in the area's other major industry: ecotourism. By re-figuring what they wanted (wilderness preservation) as a commodity (recreational area) in high demand in the global economy, they attempted to show that their own desires were in accord with the rationality of the exchange economy. They also figured themselves a more authentic embodiment of the feminine values espoused by the loggers' wives. However, they were rebuffed by numerous native leaders, who implied that the environmentalists were new-age colonialists, bent on imposing their own visions of nature, economy, and human identity. The renewal of aboriginal economies and cultures would involve a relation to the land that was different from what either the environmentalists or the loggers imagined. ... And so, on it went.

Two things are worth noticing. The first is that even those identities that apparently have biological roots—like gender and aboriginal descent—were "in play," in the sense that both the characterization and the relative priority of those identities were matters of political dispute. There is more than one way of being a woman or a native person, and (as Nancy Scott noted) there is more than one way of being a logger. The struggle itself forced people to articulate different ideas about what it meant to be this or that, or to belong "here" rather than "there." The second noticeable feature is that the effort to articulate something other than what was on offer from the company and the government entailed the specification of identities that were radically at odds with the state and the market. People were saying, in effect, "If this is what ensues from the identities you expect me to have, then I must be other than what you take me to be. I can only be what I am as an individual if I reject the form of individuality that you are thrusting upon me." To the extent that this was occurring, something more threatening than "globalization" was happening. People were actually refusing to be individuals of the sort required. ... And, of course, Clayoquot is not the only place where this sort of thing has happened.

The Discourse of Sovereignty

Post-marxism seems to have led to the crudest sort of economic determinism in social science. Everything gets attributed to the advance of the market, and the market is figured as a kind of *deus ex machina* (god outside the machine). As various analysts have argued, economic globalization is as much the result of political efforts to free business from state control, as of the logic of economic competition itself. Global markets had to be constructed, and governments were crucial to that enterprise. They were not always thinking of the greatest benefits for the greatest number of people. To put it in simplistic Marxist terms: There was a class war, and the bourgeoisie won. Whether we accept such an account or not, it reminds us of the rather obvious fact that economics is not actually separate from politics, and hence that the putative decline of national sovereignty is not to be understood as a phenomenon "dictated" by the market.

The Clayoquot example, to which I have been referring, seems to be illustrative of the decline of the nation-state. The Province of British Columbia ostensibly had authority under the Canadian Constitution to deal with the land issues at Clayoquot. But, it could not resolve those issues with the authority it had, because the market in which it was intervening was global, the media that informed the public was also global, the environmentalists with which it was contending were organized on a global scale, and the relevant public was not the provincial electorate, but the global consumerate. The political space in which the government had to operate was not a space that it could control like a sovereign. And yet, sovereignty did not pass anywhere else, at least not in any obvious way. The logging company did not assume sovereignty, nor did the media, nor did the environmentalists. If "consumers" were the ultimate sovereign, it was not very obvious to them. The most apparent feature of the situation is that no one could quite control it. And this, of course, is typical of any political situation.

So, does it follow that we have arrived at an age "beyond sovereignty"? Hardly. Sovereignty remains the dominant ideology, in that it denotes what individuals and communities are supposed to be: namely, autonomous or self-governing. As the dominant ideology, it continues to inform people's aspirations about what they should be and how they should live. Hence, notions of popular sovereignty, consumer sovereignty, individual sovereignty, national sovereignty, and, of course, state sovereignty. That states have been having some difficulty—that they no longer *appear* sovereign to many of their subjects —is obvious enough; but that is not evidence of the decline of sovereignty so much as of its transfiguration. The form of sovereignty is getting more complicated, and aspirations to sovereignty more diverse. The sovereignty to which the textbooks refer was never achieved; it was only ever an aspiration and a regulative ideal. (The discussions about Kosovo are a good indication of the ways in which that ideal can be inflected.) It is clear that conditions are changing and that some of those conditions can be understood under the rubric of globalization. It is not to be supposed, however, that the aspiration to sovereignty—which has been with us for so long—is suddenly going to disappear. What is likely, is that it will take new forms.

What I find hopeful about struggles like the ones at Clayoquot is precisely that they tend to defeat aspirations toward sovereignty. The Hayekians and other neo-liberals would like to make the market sovereign, and put it beyond politics. However, every indication is that that will not be permitted: that people will enter the market to pursue their political ends and in so doing impose terms on exchange relations that cannot be rationalized in terms of the logic of those relations. So far, so good. Other people would like to fix our identities in manageable forms, so that we can all be marketed and governed more effectively. There are many indications, however, that these forms do not fit us, and that we will struggle to throw them off. So much the better. Whether, in the process, we will learn to stop wishing that we were sovereigns, I cannot tell. But, I tend to think that experiences like those at Clayoquot are the ones most likely to loosen the hold of sovereignty thinking.

Plate 9 A symbolic entrance to Men Wu Temple ➧

文衡聖帝

8

The Mental Health of Asian Immigrants in Canada

Zheng Wu
Associate Professor, Department of Sociology, University of Victoria
Violet Kaspar
Centre for Addiction and Mental Health Foundation, Toronto

Asian Immigrants in Canada

Over the last three decades, approximately 100,000 to 250,000 immigrants arrived in Canada each year. According to the 1996 Census, Canada was home to approximately 5 million immigrants, representing a 14.5% increase since 1991. In 1996, foreign-born Canadians constituted 17.4% of the population, the largest share in more than 50 years (Canadian Census, 1996).

Along with the increase in the number of immigrants, the source countries from which Canada admits immigrants have also shifted in recent years. Throughout much of the nation's history, the majority of immigrants arrived from "traditional" source countries such as the United Kingdom, Western Europe, and the United States. The 1960s saw Canada's immigration policy come under attack as discriminatory on the basis of national origin. The Canadian federal government lifted the source country restriction in 1962, and introduced the point system in 1967. The 1976 Immigration Act reinforced the non-discriminatory policy towards independent immigrants, family reunification, and refugee issues (Beaujot, 1991). Consequently, since the mid 1960s, Canada has been admitting increasing numbers of immigrants from "non-traditional" source countries in Asia and the Pacific Islands, South and Central America, Africa, and the Middle East.

This decade has seen some of the most remarkable demographic shifts in trends of immigration to Canada. During this period, more than half of all immigrants to Canada came from Asian countries. The share of Asian immigrants as a percentage of all immigrants to Canada increased from 1-3% in the 1950s to over 60% in the late 1990s (Employment and Immigration of Canada, 1999). For example, in 1996-97, Asian immigrants constituted 63% of all immigrants landing in Canada.

The rapid influx and diversity of immigrants to Canada raise important concerns about the impact of migration on the economic, and physical and psychological wellbeing of newcomers. In this article, we use longitudinal data from the National Population Health Survey (NPHS) to examine the role of selection and migration factors in predicting successful resettlement and wellbeing of immigrants. As Asians represent one of the largest and fastest growing groups of newcomers to Canada, we focus on the mental health of Asian immigrants relative to native-born Asians and non-Asians (immigrant and non-immigrant).

Conceptual Framework

A considerable volume of research has been devoted to the study of the processes by which psychosocial stress contributes to adaptive and maladaptive social and psychological functioning (e.g., Aneshensel, 1992; Wheaton, 1996). The *stress process* model is, arguably, the most influential conceptual framework guiding the study of psychosocial influences on mental health (Pearlin et al., 1981). It focuses on the social etiology of distress/disorder by linking stressful life events and onset of distress and mental disorder. The model illustrates the interconnections between life stressors, social and psychological resources, and distress and disorder (mental health outcomes). It emphasizes the manifestation of stressors as health or adjustment outcomes, and the mediating (buffering) effects of resources in the stress process. It assumes that the presence of, or the degree of utilizing, social and psychological resources may reduce the direct impact of stressors on mental and physical health (e.g., Aneshensel et al., 1991; Lin and Ensel, 1989).

The stress model has received wide empirical support both in and outside North America. Notwithstanding the well-articulated and empirically substantiated premises regarding the processes by which individuals cope with stress, and/or succumb to the effects of stress, the existing literature is based largely on research conducted in Western cultures (e.g., Kleinman, 1987; Littlewood, 1990). Given that approximately 70% of the world population resides in what are considered collectivistic cultures (e.g., Triandis, 1990), a literature based on the experiences of members of individualistic cultures poses a serious conceptual problem for the study of stress and coping processes. However, several recent studies have shown that the stress process paradigm can be successfully adapted to studies of immigrant populations, particularly Asian immigrants (e.g., Beiser and Hyman, 1997; Noh and Avison, 1996).

From the *stress process* perspective, we evaluate three competing hypotheses pertaining to mental health differences between immigrant and non-immigrant populations, and between Asian and non-Asian populations. To eliminate confounding effects of other risk factors, our analysis incorporates influences of socially induced stressors (e.g., stressful life events) and social and psychological resources (e.g., social support, sense of coherence, self-esteem) stipulated in the stress process framework (Pearlin et al., 1981).

Hypotheses

The Stress-Distress Hypothesis

Voluntary migration involves an uprooting process that disrupts social networks, imposing a heavy strain on the mental health of immigrants (Beiser et al., 1988a). Psychological distress may result from the undesirable conditions surrounding exodus, including pre-migration stress, family composition, and unrealistic expectations of the future. For example, the decision to emigrate is often imposed by financial constraints that supersede the considerations of women and children in the family. Studies have shown that single migrants, married migrants whose spouses and children have been left behind, and parents who have been separated from their young children tend to experience elevated risks of mental disorder (e.g., Burke, 1980).

Public attitudes toward immigrants may also affect mental health in new *and* old immigrants. Even in countries that encourage immigration for economic reasons, public attitudes toward immigrants tend to be ambivalent (Littlewood and Lipsedge, 1989). While their economic contributions are welcome, newcomers are expected to conform to the customs of the host society. The contribution of their labour is doubted when the (growth of) economy in the host country declines and labour supply increases. In Canada, unlike the US, the official ideology of multiculturalism should accord well with the mental health of newcomers, especially immigrants from non-European countries. However, while people from other cultures are encouraged to "embrace a new culture, to preserve their own or to combine the two," a significant gap remains between ideal public policies and actual behaviour in daily life (Beiser et al., 1988a, p. 11).

While most immigrants may feel the stresses surrounding migration, emotional difficulties following migration tend to increase when cultural differences between the home and the host society are accentuated (Danna, 1980). Common sense would tell us that some familiarity with the host culture(s) and the host language(s) should facilitate the adjustment to the new society. Lack of knowledge of either official language may raise the stress of immigrants, particularly newly arrived immigrants. In fact, recent research has shown that inability to speak the host language is associated with a variety of psychological disorders (e.g., Williams and Carmichael, 1985).

Accordingly, immigrants may be more likely to succumb to poor mental health due to the stress induced by cumulative negative circumstances inherent in the process of migration. Despite the intuitive appeal of this dose-response relationship, recent findings suggest that the level of stress may decline with, for example, the length of residence, realistic expectations of the future, and fluency in the host language. In this chapter, we test the hypothesis that immigrants exhibit higher risks of mental health problems than non-immigrants. This process reflects the classic "*stress equals distress*" hypothesis (Selye, 1956). We will refer to this as the *stress-distress hypothesis.*

The Self-Selection Hypothesis

Migration is selective. While reasons for migration vary from one individual to another, the desire to improve material conditions (or political conditions, as in the case of refugee migration) appear to be the primary motive behind the movement across national boundaries (Rystad, 1992). Theories of voluntary migration suggest that migration tends to select individuals who have more human (social) and/or economic resources (e.g., Lucas, 1981; Salt, 1992). For example, research has shown that international (and internal) migration is more common among men, younger and better educated people, and people with more financial resources (Rystad, 1992). Does migration also select people with better, or worse, health status?

At one time, it was popular belief that some groups of immigrants experienced higher risks of mental disorders because they were more liable to certain types of mental illness. Indeed, it was suggested that some European countries deliberately sent people who had mental illness (Littlewood and Lipsedge, 1992). By 1882, mentally ill people were prohibited from emigrating to the US, and immigrants who were detected to be mentally ill within 1 year of arrival were subject to deportation. By the beginning of this century, the ban on mentally ill migrants was extended to include alcoholics and illiterates over the age of 15.

While international migration may not select people who have mental problems, does it attract people who have better mental health? We suspect that the answer may be yes. A common assumption is that voluntary migration may well select individuals who are more resilient, determined, and willing to take risks. They are considered better prepared for life challenges, more equipped for life changes, and more adaptable to new environments. In short, voluntary immigrants may share a so-called "hardy" personality, which can be manifested as greater self-control, self-confidence, coping skills, and the ability to explore and search for social resources when they face distressful situations (Kobasa and Maddi, 1977; Kuo and Tsai, 1986). Therefore, it may be reasonable to assume that immigrants may well experience better psychological health to begin with than non-immigrants partly because of their potential to mobilize more psychological resources for battling loneliness, alienation, and, in some cases, drops in social status.

Notwithstanding, migration, whether it is internal or cross-border migration, is costly. While economic costs (e.g., travelling to the destination and living expenses while seeking employment) are obvious, there are also human costs. Psychological strains of being alone in a new country, away from family and friends for extended periods of time, are likely to increase emotional difficulties in adapting to the new environment. In fact, as Lucas (1981) has shown, high psychological costs are the single most important factor explaining the phenomenon of immigrants returning to their country of origin. In this study, we test the hypothesis that there is a selection effect in the assessment of mental health of immigrants, and that the difference in the risk of mental problems between the two groups is due to the (positive) selection into migration. We will call this the *selection hypothesis*.

The Ethno-Cultural Hypothesis

The third hypothesis that we evaluate in this study pertains to the role of culture and ethnicity in mental health. We test the hypothesis that Asian immigrants tend to experience lower risks of mental health problems than non-Asian immigrants in Canada. We will call this the *ethno-cultural hypothesis*.

Empirical evidence has been consistent with the idea that Asian immigrants enjoy better mental health than other ethnic/racial groups (e.g., Littlewood and Lipsedge, 1989; Noh and Avison, 1996). However, the findings, in particular the lower rate of hospital admissions for mental health problems among Asian immigrants (e.g., Jones and Korchin, 1982; Kitano, 1982), may reflect language difficulties and, more importantly, cultural tendencies toward under-utilization of mental health services and the stigma attached to mental illness in many Asian societies (e.g., Takeuchi et al., 1988). While under-utilization of mental health services may reflect language difficulties and cultural differences between Asians and non-Asians, two other explanations are also plausible.

One is the Asian preference of responding to environmental demands through indirection and self-regulation rather than direct change or management of stressful situations (see, e.g., Noh et al., 1999). It is reminiscent of perception-focused coping which entails a cognitive recasting of stressful events that may be particularly effective in dealing with stressors that are less amenable to personal control (Pearlin and Schooler, 1978; Wethington and Kessler, 1991), but are associated with unfavourable outcome expectancies (Abella and Heslin, 1989). Faced with chronic and negative life experiences, including those associated with resettlement, one may cope by adopting such beliefs as "adversity makes one a better person." Even in some inventories designed to assess coping in Asian studies, researchers have constructed items such as "it is best to do nothing" and "to lose is to win," which are reflective of Asian beliefs and proverbs (e.g., Furukawa et al., 1993).

It is also a possibility that Asian immigrants tend to have better mental health because they have more realistic expectations of a Canadian sociocultural milieu that will be largely unfamiliar to them. Thus, they may arrive psychologically prepared to face the hardships associated with resettlement (e.g., Littlewood and Lipsedge, 1989). Such anticipatory perceptions of cultural and lifestyle variations between East and West may be protective, and effective in coping with stressors that are associated with migration and settlement.

Data and Methods

Data

Our study draws longitudinal data from the National Population Health Survey (NPHS), conducted by Statistics Canada. The first cycle of data collection took place between June 1994 and June 1995. The survey is expected to continue every 2 years for a period of two decades. The 1994-95 NPHS used a national probability sample of 17,626 Canadian residents aged 12 years and older, excluding residents of the Yukon and Northwest Territories, people on Indian reserves, and full-time institutionalized residents. Telephone interviews were used to collect the data. The survey collected detailed information on health status, health risk factors, socioeconomic status, and demographic and family characteristics. The overall response rate for the 1994-95 NPHS was 84.8% (see Statistics Canada, 1996, 1998).

Of the 17,626 respondents included in the first cycle of the NPHS, 14,786 (due to sample attrition) and an additional 2,490 respondents (mainly those under the age of 12 in 1994-95 and now eligible for the NPHS program) were eligible for the longitudinal panel. The overall response rate for the longitudinal panel was 93.6%. Of the 16,168 respondents who agreed to participate, 15,670 provided complete health information and the remaining 498 provided partial information. This study uses data from those who provided complete information for both cycles of the survey. After deleting missing cases for key (psychometric) variables, the study sample includes 11,653 respondents.

Measures

In this study, we consider two mental health indicators. First is a dichotomous variable indicating whether the respondent had experienced a major depressive episode (MDE) in the last 12 months. MDE is derived from a shortened version of the Composite International Diagnostic Interview (CIDI) developed by researchers at the University of Michigan (e.g., Wade and Cairney, 1997). The instrument was designed to measure 1-year population prevalence rates of diagnosable depression. The second measure of mental health is a numerical scale of depression. The scale is based on a subset of items from the CIDI, and is available in the NPHS (Statistics Canada, 1996).

Our primary independent variable is immigrant/ethnic status. It is measured as a 4-level categorical variable: a) Asian immigrants; b) Asian non-immigrants; c) non-Asian immigrants; and d) non-Asian and non-immigrants. Level 4, the non-Asian and non-immigrant population, is used as the reference category in comparisons.

To control for the confounding effects of other health determinants, we include several observed individual-level characteristics that are known to be important (e.g., Aneshensel et al., 1991; Lin and Ensel, 1989; Noh et al., 1999). The selection of such control variables not only depends on their theoretical significance, but is necessarily constrained by the NPHS data. For these reasons, we consider four groups of control variables: 1) health risk factors; 2) psychological and social resources; 3) socioeconomic status (e.g., social class, education, and income); and 4) demographic characteristics (e.g., gender, age, marital status, and census metropolitan area). To conserve space, variable definitions and descriptive statistics for these variables are not provided in the chapter, but are available from the authors upon request.

Methods

Our empirical analysis proceeds in two steps, starting with a series of multiple comparisons to contrast the mental health of Asian immigrants with people in other ethnic/immigrant groups. These comparisons use ethnic/immigrant status as the only explanatory variable and lead to an overall qualitative assessment of the relative mental health of Asian immigrants as a group vs the other three groups (non-Asian immigrants, Asian non-immigrants, and non-Asian and non-immigrants). To evaluate the net impact of ethnic/immigrant status we then add other potentially relevant explanatory variables to the model. The analysis will help us evaluate the *stress-distress* and *ethno-cultural hypotheses*.

In step 2, we test the *selection hypothesis*. We first model the process of self-selection into international migration. Biases in the regression estimates introduced through self-selection have been extensively discussed in the literature, and several procedures have been proposed to correct selectivity (e.g., Greene, 1993). To model the selection into migration, we followed a procedure suggested by Heckman (1976, 1979). We first computed a probit model of the propensity to migrate, which is then used to estimate selectivity into migration (the probit estimates are available from the authors). We then repeat the analysis in step 1 with the estimated selection factor added to the model. The significance of the selection factor, the hazard rate, will indicate whether the effect of selectivity is relevant in the assessment of mental health. The analysis will also confirm whether the observed effect of ethnic/immigrant status, if there is any, is due to the self-selection into migration.

As our regression analysis involves one dichotomous and one continuous dependent variable, we employ generalized linear model (GLM) techniques in the data analysis (McCullagh and Nelder, 1989).

Results

Trends in Asian Migration to Canada

To set the stage for analysis, we begin with a brief look at the trends of Asian immigrants to Canada in recent decades. Figure 8.1 shows the number of Asian immigrants to Canada between 1955 and 1998. For the purposes of comparison, the number of all immigrants to Canada is also provided in the figure. Figure 8.2 plots the number of Asian immigrants as a percentage of all immigrants to Canada over the same time period. Overall, there has been a clear, rising trend of Asian immigration to Canada both in an absolute and a relative sense. For example, in 1955, less than 4,000 Asian immigrants came to Canada. This number rose to 47,000 in 1975, and over 100,000 in the 1990s. The number of Asian immigrants peaked in 1993, when nearly 150,000 Asian immigrants landed in Canada.

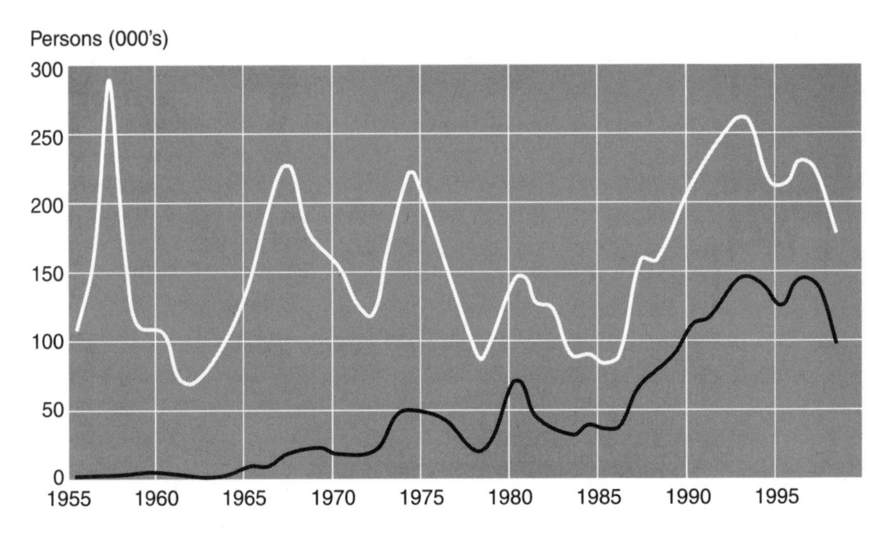

Figure 8.1 Number of immigrants to Canada, 1955-1998

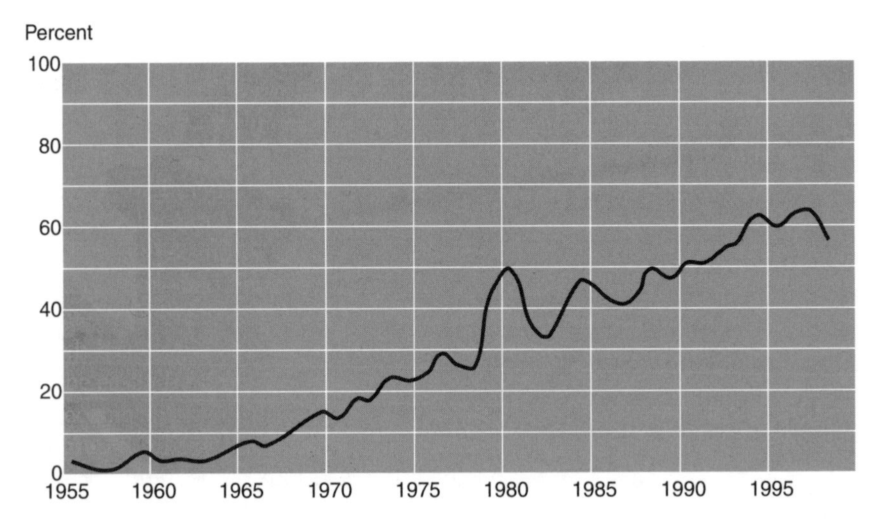

Figure 8.2 Asian immigrants as a percentage of all immigrants to Canada, 1955-1998

The relative numbers tell a similar story. In 1955, of the 112,000 immigrants who came to Canada, only 3% came from Asian countries. This percentage rose to 25% in 1975, 45% in 1985, and has been over 50% in the 1990s. In 1996-97, the percentage of Asian immigrants went up to 63%. The low number of Asian immigrants prior to the early 1960s largely reflects Canada's discriminatory immigration policy, which gave preference to immigrants from Britain, the US, and Western Europe. The large inflows of immigrants from less developed countries began in the late 1960s, when the Canadian federal government adopted a nondiscriminatory policy by establishing a "point system" for selecting independent immigrants.

Mental Health of Asian Immigrants

Do Asian immigrants enjoy better mental health than others? Table 8.1 presents the GLM parameter estimates for the two mental health indicators: depression and MDE (major depressive episode). Panel A in the table shows the multiple linear contrasts comparing Asian immigrants with other ethnic/immigrant groups. Panel B adds control variables to the model.

Panel A shows that Asian immigrants have a significantly lower risk of depression (measured at the second cycle of the NPHS) than the general population (the "non-Asian and non-immigrant" population; hereafter NA-NI). They are, however, not significantly different from either "Asian non-immigrants" (hereafter A-NI) or "non-Asian immigrants" (hereafter NA-I). When controls are added to the model (see panel B), the negative effect of Asian immigrant status remains significant. Further, the results of the linear contrasts suggest that "Asian immigrants" (hereafter A-I) also have a lower risk of depression than NA-I. These results provide some support for the *ethno-cultural hypothesis* that Asians, particularly Asian immigrants, tend to have better mental health than others. However, the findings that A-I do not differ significantly from A-NI, and that NA-I do not differ significantly from NA-NI (the reference group) run against the *stress-distress hypothesis*, suggesting that immigration in and of itself does not generate significant stresses that negatively influence the mental health of immigrants.

While the results are not shown in tables, we also ran GLM models separately for immigrant and non-immigrant groups. Through restricting our analysis to the immigrant population, we found additional support for the *ethno-cultural hypothesis*—Asian immigrants appear to experience a lower risk of depression, even after controlling for length of residence in Canada and other risk factors (e.g., negative life events, resources, SES, and gender). However, the effects are only marginally significant ($p < .10$). Using the same model specifications for non-immigrants, we found that Asian Canadians do not have significantly lower risks of depression than non-Asian Canadians, although the signs of the estimates were congruent with the *ethno-cultural hypothesis*.

Table 8.1 Generalized linear models of mental health (T2) on ethnicity and immigration status, Canada: 1994-1997

Independent Variable	*Depression (T2)* (Normal)	*MDE (T2)* (Binomial)
A. Without Controls		
Asian immigrant	-0.177 **	-0.821 **
Asian non-immigrant	-0.106	-0.084
Non-Asian immigrant	-0.049	-0.195
Non-Asian and non-immigrant[a]		
Linear Contrast (LR c^2)		
Asian immigrant vs Asian non-immigrant	0.110	0.692
Asian immigrant vs non-Asian immigrant	2.104	2.375
Intercept	0.341 ***	-3.030 ***
Log Likelihood	-20077	-2122
B. With Controls[b]		
Asian immigrant	-0.211 ***	-1.104 **
Asian non-immigrant	-0.114	-0.143
Non-Asian immigrant	-0.051	-0.192
Non-Asian and non-immigrant[a]		
Linear Contrast (LR c^2)		
Asian immigrant vs Asian non-immigrant	0.228	1.048
Asian immigrant vs non-Asian immigrant	3.571 *	4.600 **
Intercept	1.622 ***	0.632
Log Likelihood	-19423	-1754

[a] Reference category

[b] Model controls for a number of explanatory variables (not shown in the table)

* $p < .10$ ** $p < .05$ *** $p < .01$ (two-tailed test)

Selection into Migration

In Table 8.2, we present the results of selection models of depression. It shows some evidence of self-selection into migration in the models of depression as the selection factor, the hazard rate, is statistically significant (see panel A). However, when controls are added, the effect of the selection factor is no longer significant. While previously observed effects associated with Asian immigrants are nonsignificant in uncontrolled models, these effects become significant when controls are added. These findings provide evidence that there is no meaningful selection effect in the assessment of immigrant mental health. The reduced risk of depression observed in Asian immigrants is not a consequence of self-selection into migration, which is inconsistent with the *selection hypothesis*.

Table 8.2 Generalized linear models of mental health (T2) on ethnicity and immigration status controlling for selection to international migration, Canada: 1994-1997

Independent Variable	*Depression (T2)* (Normal)	*MDE (T2)* (Binomial)
A. Without Controls		
Asian immigrant	0.133	0.555
Asian non-immigrant	0.183	1.197
Non-Asian immigrant	-0.025	-0.092
Non-Asian and non-immigrant[a]		
Linear Contrast (LR c^2)		
Asian immigrant vs Asian non-immigrant	0.056	0.532
Asian immigrant vs non-Asian immigrant	1.093	0.982
λ (hazard rate)	-0.208 **	-0.966 ***
Intercept	0.381 ***	-2.847 ***
Log Likelihood	-20074	-2118
B. With Controls[b]		
Asian immigrant	-0.388 **	-1.735 **
Asian non-immigrant	-0.283	-0.720
Non-Asian immigrant	-0.062	-0.229
Non-Asian and non-immigrant[a]		
Linear Contrast (LR c^2)		
Asian immigrant vs Asian non-immigrant	0.274	1.166
Asian immigrant vs non-Asian immigrant	4.230 **	4.131 **
λ (hazard rate)	0.119	0.417
Intercept	1.639 ***	0.211
Log Likelihood	-19422	-1754

[a] Reference category

[b] Model controls for a number of explanatory variables (not shown in the table)

* $p < .10$ ** $p < .05$ *** $p < .01$ (two-tailed test)

Discussion

On July 20, 1999, a group of 123 Chinese migrants were spotted near Gold River off the west coast of Vancouver Island, British Columbia, in a rusty freighter. They were immediately detained by the Canadian authorities. It was reported that they had spent 38 days crossing the Pacific Ocean. Most of the Chinese migrants subsequently claimed refugee status. Later, some 83 migrants were released into the community to await hearings of their refugee status claims.

The other migrants remained in custody under investigation for human smuggling charges. With no lifeboats on board, these migrants risked their lives in the quest to better themselves politically and/or economically. Whether or not these migrants are really political refugees is yet to be determined. However, their experience is clearly consistent with the view that only the most determined, resilient, and desperate are likely to emigrate (Rystad, 1992).

In this study, we have examined the psychological wellbeing of Asian immigrants by comparing their mental health status with that of other immigrant and non-immigrant populations. Because leaving one's home country and settling in a new society are critical and stressful life events, we hypothesized that the stress associated with the process of migration may raise the risk of mental health problems in the immigrant population. Accordingly, the risk of depression may be expected to be higher among immigrants than non-immigrants. However, our analysis provided no support for this hypothesis. Within the Asian sample, we found no evidence that immigrants are any less comfortable psychologically than non-immigrants. We found no significant mental health differences between immigrants and non-immigrants within the non-Asian sample. This pattern of results lends no support to the *stress-distress hypothesis*, and suggests that immigration in and of itself does not threaten the mental health of immigrants. A similar conclusion is presented in a recent task force report on immigrant mental health that "migration *per se* does not predict an increased risk of mental disorder" (Beiser et al., 1988b, p. i). The results of Canadian studies of Korean immigrants in Toronto (Noh and Avison, 1996) and Southeast Asian refugees in Vancouver (Beiser and Fleming, 1986) also suggest that migrants are not necessarily at increased risk for psychological distress symptoms.

Voluntary migration is selective of individuals who are the most determined and resilient. For this and other reasons (e.g., all legal immigrants are required to take a complete physical examination before entering Canada), we hypothesized that international migration tends to select individuals who may be mentally healthier. Our analysis lends little support to this hypothesis. We found that selectivity into migration has no meaningful bearing on the assessment of immigrant mental health. However, this finding does not reject the notion that migration is selective, as our data have shown that the propensity to emigrate varies according to several personal characteristics such as age and socioeconomic status (detailed results are available from the authors). What it means is that migration does not necessarily select individuals who are mentally the "fittest."

British researchers Littlewood and Lipsedge (1989) argued that it would be fallacious to talk of immigrants (to Britain) as a single, homogeneous group of people when it comes to mental health. Immigrants differ, to a large degree, depending upon the country of origin. The results of our analysis support the hypothesis that Asian immigrants enjoy better mental health than other immigrants or non-immigrant Canadians.

Clearly, there is something about Asians, particularly Asian immigrants, that makes them less susceptible to stress. As we argued, the mental health advantage of Asians may reflect cultural differences reflected in their responses to stressful circumstances (e.g., Noh et al., 1999). Alternatively, cultural prohibitions may discourage Asians from seeking professional help for symptoms of psychological distress (e.g., Takeuchi et al., 1988). The psychological preparedness for experiences of resettlement-related stresses, and the support of established community networks offer other interpretations for the observed mental health advantages of Asian immigrants (e.g., Noh et al., 1994). While NPHS data are not amenable to investigating the role of like-ethnic social support, visible residential segregation of Asian communities in Canadian metropolises provides indirect evidence that settlement in an ethnic enclave, with its provision of access to ethnic networks and resources (social and psychological) may have played a vital role in meeting Asian immigrants' instrumental and emotional needs in both routine and crisis situations (Kuo and Tsai, 1986; Noh and Avison, 1996). These ideas remain empirical questions that should be further examined in future research.

References

Abella, R., and Heslin, R. (1989). Appraisal processes, coping, and the regulation of stress-related emotions in a college examination. *Basic and Applied Social Psychology*, 10, 311-327.

Aneshensel, C. S., Rutter, C. M., and Lachenbruch, P. A. (1991). Social structure, stress, and mental health: Competing conceptual and analytic models. *American Sociological Review*, 56, 166-178.

Aneshensel, C. S. (1992). Social stress: Theory and research. *Annual Review of Sociology*, 8, 15-38.

Beaujot, R. (1991). *Population change in Canada: The challenges of policy adaptation*. Toronto: McClelland and Stewart.

Beiser, M., et al. (1988a). *After the door has been opened: Mental health issues affecting immigrants and refugees in Canada*. Canadian Task Force on Mental Health Issues Affecting Immigrants and Refugees. Ottawa: Ministry of Supply and Services of Canada.

Beiser, M., et al. (1988b). *Review of the literature on migrant mental health*. Canadian Task Force on Mental Health Issues Affecting Immigrants and Refugees. Ottawa: Ministry of Supply and Services of Canada.

Beiser, M., and Fleming, J. A. (1986). Measuring psychiatric disorder among Southeast Asian refugees. *Psychological Medicine*, 16, 627-639.

Beiser, M., and Hyman, I. (1997). Refugees' time perspective and mental health. *American Journal of Psychiatry*, 154(7), 996-1002.

Burke, A. W. (1980). Family stress and the precipitation of psychiatric disorder: A comparative study among immigrant West Indian and native British patients in Birmingham. *International Journal of Social Psychiatry*, 26(1), 35-40.

Danna, J. J. (1980). Migration and mental illness: What role do traditional childhood socialization practices play? *Cultural and Medical Psychiatry*, 4(1), 25-42.

Employment and Immigration of Canada (1999). *Immigration to Canada by country of last permanent residence*. SDDS 3601. 26 June.

Furukawa, T., Suzuki-Moor, A., Saito, Y., and Hamanaka, T. (1993). Reliability and validity of the Japanese version of the coping inventory for stressful situations (CISS): A contribution to cross-cultural studies of coping. *Seishin Shinkeigaku Zasshi*, 95, 602-620.

Greene, W. H. (1993). *Econometric analysis*. Second edition. New York: Macmillan.

Heckman, J. (1976). The common structure of statistical models of truncation, sample selection and limited dependent variable and a simple estimator for such models. *Annals of Economic and Social Measurement*, 5, 475-492.

Heckman, J. (1979). Sample selection bias as specification error. *Econometrica*, 47, 153-161.

Jones, E. E., and Korchin, S. J. (1982). *Minority mental health*. New York: Praeger.

Kitano, H. H. (1982). Mental health in the Japanese-American community. In E. E. Jones and S. J. Korchin (Eds.), *Minority mental health* (pp. 149-164). New York: Praeger.

Kleinman, A. (1987). Anthropology and psychiatry: The role of culture in cross-cultural research on illness. *British Journal of Psychiatry*, 151, 447-454.

Kobasa, S. C., and Maddi, S. R. (1977). Existential personality theory. In R. J. Corsini (Ed.), *Current personality theories* (pp. 243-76). Itasca, Il: F. E. Peacock.

Kuo, W. H. and Tsai, Y. (1986). Social networking, hardiness and immigrant mental health. *Journal of Health and Social Behavior*, 27, 133-149.

Lin, N., and Ensel, W. M. (1989). The stress and health: Stressors and resources. *American Sociological Review*, 54, 382-399.

Littlewood, R. (1990). From categories to contexts: A decade of the "new cross-cultural psychiatry." *British Journal of Psychiatry*, 156, 308-327.

Littlewood, R., and Lipsedge, M. (1989). *Aliens and alienists: Ethnic minorities and psychiatry*. Second edition. London: Unwin Hyman.

Lucas, R. E. B. (1981). International migration: Economic causes, consequences and evaluation. In M. M. Kritz, C. B. Keely, et al. (Eds.), *Global trends in migration: Theory and research on international population movements* (pp. 84-109). Staten Island, New York: Center for Migration Studies.

McCullagh, P., and Nelder, J. A. (1989). *Generalized linear models*. Second edition. London: Chapman and Hall.

Noh, S., and Avison, W. R. (1996). Asian immigrants and the stress process: A study of Koreans in Canada. *Journal of Health and Social Behavior*, 37, 192-206.

Noh, S., Wu, Z., and Avison, W. R. (1994). Social support and quality of life: Sociocultural similarity and the efficacy of social support. In G. Albrecht and R. Fitzpatrick (Eds.), *Advances in medical sociology, Vol. V: Quality of life in health care* (pp. 115-137). Greenwich, CT.: JAI Press.

Noh, S., Beiser, M., Kaspar, V., Hou, F., and Rummens, A. (1999). Perceived racial discrimination, depression, and coping: A study of Southeast Asian refugees in Canada. Forthcoming in *Journal of Health and Social Behavior*, 193-207.

Pearlin, L. I., Lieberman, M. A., Menaghan, E. G., and Mullan, J. T. (1981). The stress process. *Journal of Health and Social Behavior*, 22, 337-56.

Pearlin, L. I., and Schooler, C. (1978). The structure of coping. *Journal of Health and Social Behavior*, 19, 2-21.

Rystad, G. (1992). Immigration history and the future of international migration. *International Migration Review*, 26(4), 1168-99.

Salt, J. (1992). The future of international labor migration. *International Migration Review*, 26(4), 1077-1111.

Selye, H. (1956). *The stress of life*. New York: McGraw-Hill.

Statistics Canada (1996). *National Population Health Survey Public Use Microdata File User Documentation*. Ottawa: Ministry of Supply and Services.

Statistics Canada (1998). *1996-97 National Population Health Survey Public Use Microdata File User Documentation*. Ottawa: Ministry of Supply and Services.

Takeuchi, D., Phil, T., Leaf, J., and Kuo, H.S. (1988). Ethnic differences in the perception of barriers to help-seeking. *Social Psychiatry and Psychiatric Epidemiology*, 23, 273-280.

Triandis, H. (1990). Cross-cultural studies of individualism and collectivism. In J. J. Berman (Ed.), *Nebraska symposium on motivation, 1989: Cross-cultural perspectives* (pp. 41-133). Lincoln, NE: University of Nebraska Press.

Wade, T. J., and Cairney, J. (1997). Age and depression in a nationally representative sample of Canadians: A preliminary look at the National Population Health Survey. *Canadian Journal of Public Health*, 88, 297-302.

Wethington, E., and Kessler, R. (1991). Situations and processes of coping. In J. Eckenrode (Ed.), *The social context of coping* (pp. 13-29). New York: Plenum Press.

Wheaton, B. (1996). The domains and boundaries of stress concepts. In H. B. Kaplan (Ed.), *Psychosocial stress: Perspectives on structure, theory, life-course, and methods* (pp. 29-70). San Diego, CA: Academic Press.

Williams, H., and Carmichael, A. (1985). Depression in mothers in a multi-ethnic urban industrial municipality in Melbourne. Aetiological factors and effects on infants and preschool children. *Journal of Child Psychology and Psychiatry*, 26(2), 277-288.

Plate 10 Smoking Buddha, Taiwan ➧

Section II

Economic Growth and Development in the Asia-Pacific Region

9

Economic and Demographic Integration in the Asia-Pacific and Structural Change in Japan and Pacific Canada[1]

Carl Mosk

Professor, Department of Economics, University of Victoria

Introduction

Over the course of the last half century, the Asia-Pacific region has become increasingly integrated. By "integration" I mean the intertwining of the economic and demographic interests of individual countries within the region. Evidence of increasing integration is provided in Table 9.1.

The strengthening of integration over the last five decades is largely due to a set of forces simultaneously promoting divergence and convergence in the Asia-Pacific. Convergence refers to a narrowing of gaps between countries in average standard of living, knowledge of technology, structure of production activities, and per capita levels of human and physical capital. Divergence is the widening of gaps in these factors. In this chapter I argue three propositions. The first is that there are seven key forces conditioning convergence and divergence: investment in physical capital, education and skill formation, diffusion of technology, international migration, international trade, policy making, and mega-city growth. Second, I refer to evidence that the net effect of these forces is to strengthen convergence. Third, I argue that, because the degree of convergence within the Asia-Pacific region has improved, the national economies of the region have been increasingly integrated. However while this is true at the regional level, it is not necessarily valid at the bilateral national level.

While this line of reasoning is abstract, my goal is empirical and concrete. In Section II of the chapter, I refer to findings on divergence and convergence, demonstrating that convergence has triumphed. Section III focuses on post-Cold War economic growth in the two most important economies of the region, the US and Japan. Section IV deals with demographic integration and concentrates on the contrast between "open" Pacific Canada and "closed" Japan. Picking up on this bilateral comparison, Section V discusses structural change in Pacific Canada and Japan occasioned by convergence in the Asia-Pacific. Finally, in Section VI, I summarize the key points of the chapter.

Plate 11 Garden, Taipei

Table 9.1 Companies in British Columbia and in Canada doing business with Japan: 1994/95 and 1997/98

Panel A: Number of companies in Canada and in BC doing business with Japan

1994-95			*1997-98*		
	In BC (number and percent of all companies in Canada)			*In BC (number and percent of all companies in Canada)*	
In Canada	Number	Percentage	In Canada	Number	Percentage
4151	880	21.2	4856	1151	23.7%

Panel B: Change in number of companies doing business with Japan between two periods, 1994-5/1997-8 [rates of increase are rates per 100 companies and are expressed in %]

Absolute Increase			*In BC: Rates of Increase (IR), Disappearance (DR), and Addition (AR)*[(a)]		
	In BC (number and percent of total Canadian increase)				
In Canada	Total	Percentage	IR	DR	AR
705	271	+38.4	30.8%	-14.2%	+45.0%

Panel C: Percentage of BC companies doing business with Japan in 1997-98 in 10 categories (*indicates increase in % between 1994-5 and 1997-8)

Agricultural and related	2.0
Oil and minerals	1.6
Food and food processing	10.9
Fabricated products	1.3
Forest products	8.3
Chemical products	1.3
Manufacturing	30.1*
Construction/building equipment and supplies	10.5*
Services (except trading houses and associations)	24.7*
Trading houses and associations	10.2*

[(a)] DR = rate of companies vanishing between 1994/5 and 1997/98 (&/or completely changing corporate name so that matching of old and new company names proved impossible); AR = rate of companies being newly created between 1994/5 and 1997/98 (includes comprehensive corporate name changes).

Sources: Canada-Japan Trade Council (undated (a)) (undated (b)).

Convergence and Divergence in the Asia-Pacific Region

A host of standard indicators of economic development point to convergence between Asia and North America/Western Europe.[2] Of these indicators, the most widely used is the flow of national or domestic income/product, and the level of income per capita. For instance, using GDP calculated in US dollars adjusted for purchasing power parity of currencies as a measure of production, it appears that Asia's share of world production has jumped by leaps and bounds since the 1960s. The share of Japan was 5.6% in 1967; it was 7.8% in 1989. The share of developing Asia was 11% in 1967; by 1989 this percentage had reached 19.3%. Per capita income between 1975 and 1990 also converged. Setting per capita GDP in the US at 100, South Korea jumped from 20.7 to 35.2 (% of US levels), Malaysia from 21.5 to 26, and Thailand from 13 to 19.3. To be sure, these figures must be treated with caution. Income is an imperfect index of development. But we should not throw out the baby with the bath water. As Panel D of Table 1 in Mosk (1999b) shows, there is a strong association between economic development measured in terms of income and the quality of life as captured by figures on life expectancy, daily calorie intake, and access to sanitary drinking water.

Figures on income also show that, within Asia, Japan is the preeminent national economy. While adjustments for purchasing power parity diminish Japan's relative standing when calculated using exchange rates, the fact remains that Japanese technology, Japanese investment, and Japanese demand for goods and services produced in other Asian countries are crucial to Asian performance. In this lies Japan's pre-eminence in Asian economic affairs.

Why has convergence occurred? The literature in economics and political economy suggests a host of factors are working to reduce disparities in income and in income per capita. But, as I shall now argue, these very same factors can also promote the diametric opposite of convergence, namely divergence.

Consider the seven factors listed in Table 9.2. As can be seen, for each factor listed, both positive and negative impacts on convergence are possible. For instance, consider the impact of investment in physical capital. In low-income countries, per capita levels of human and physical capital are low. At these low per capita levels, the marginal productivity of capital—the incremental payoff in terms of income growth arising from accumulation of capital—is high. By contrast, in capital rich highly developed countries, the incremental payoff to accumulation is far lower. Thus—provided rates of investment are equal in low- and high-income countries—convergence takes place. However, this logic overlooks the fact that for many types of capital investment, the demand for skilled labour rises with the capital/labour ratio. In these cases, human capital is complementary to physical capital. Thus, investment rates in certain industries may be higher in countries with greater levels of human capital per capita. Moreover, relying on investment to shrink gaps in economic performance between countries may encourage short-term oriented

Table 9.2 Selected impacts of investment, education and training, technological diffusion, international migration, trade, economic policy making, and urbanization on international convergence or divergence

Item and some of the ways in which it is realized	*Impact on convergence*	*Impact on divergence*
Investment and Increase in Capital/Labour Ratio	Impact of rise in capital/labour ratio generally greater in lower income per capita countries	Greater capital intensity intensifies demand for higher human capital quality (complementarity in factor inputs). Investment subject to speculation, especially in case of short-term lending from abroad (promotes instability)
Education, Training and Skill Formation	Improves pay-off to physical capital accumulation and enhances import/adaptation of technology	Professional labour mobile internationally. In environments where there are few scale economies in new R & D, human capital depreciates; hard to predict future demands
Technology and Its Diffusion through Information Technology and through Multinational Enterprises and Aid Programs	Diffusion through imitation, adaptation and the spawning of hybrids increases total factor productivity growth	Barriers to diffusion are manifold. These include national security concerns, multinational enterprise proprietary secrets, etc.
International Migration	Depending on characteristics of emigrants and the number relative to the labour force, emigration may raise wages and reduce under-employment and unemployment in sending country	"Brain drain" may draw away the most talented from the developing country to countries where new technology is being created. Depends partly on nature of immigration policies in developed countries and the extent to which refugee family reunification concerns take priority in receiving nations.

Table 9.2 (continued)

Item and some of the ways in which it is realized	*Impact on convergence*	*Impact on divergence*
International Trade Expansion due to Declining Transport Costs, to International Agreements, and to Expansion in Global Income	Labour intensive manufacturing attracted to low wage countries; Exports provide a "vent for surplus" of production	Static comparative advantage and path dependence which discourages deviating from former comparative advantage may hinder development; intra-industry trade in developed countries associated with scale economies; unions and business federations in high income countries lobby against imports, secure Voluntary Export Restrictions and other protection.
Policies (Keynesian, Input-Output Planning, Trade, Population, Education)	Aggregate monetary and fiscal instruments and institutions in developed countries serve as models for similar institutions in developing countries. Refinement policies offers rich menu of options	Conflicting models for aggregate policy in Asia-Pacific: Open (North America) style policy approach vs Closed (Japan) style policy approach. Changes in international environment rapidly changing relative attractiveness of policy options.
Mega-City Growth and Urbanization in Developing World	Facilitates emigration of semi-skilled and unskilled labour to large metropolitan centers in the developed world. Mega-cities provide scale economies in technological diffusion, infrastructure for energy, labour, capital, and for transportation.	Encourages brain drain of talented individuals to large metropolitan centers in developed world; encourages pollution, environmental degradation, and spread of some infectious diseases. Crime is often fostered in large urban centers.

speculative investment. Speculation may generate instability. Developing countries may end up riding a roller coaster of boom and bust. In addition, instability may sow the seeds of chaos and rampant bankruptcy, thereby discouraging further investment. So divergence could also occur.

Now let us consider our second factor: education, training, and skill formation. On the one hand, investing in academic infrastructure improves the payoff to physical capital accumulation. Moreover, since technical personnel are required for successful reverse engineering and emulation of technological feats achieved in other countries, promoting scientific and engineering education will facilitate the import of technology from abroad. Thus, investing in human resources may promote convergence. However, such investments may also backfire. Unlike physical capital, human capital consciously responds to incentives. Professional labour is internationally mobile. As it depreciates rapidly in environments where there is a paucity of R & D, it is attracted to locales where the return on capital is potentially the highest. That is, professionals are drawn to countries where there are substantial scale economies in research, that is to North America, Western Europe, and Japan.

Our third candidate, technological diffusion through imitation, adaptation, and the hiving off of hybrids, is a potentially powerful impetus to convergence. Constraining this factor are strong barriers to the diffusion of technology. National security and corporate concerns over the dissemination of proprietary innovation are two prime barriers.

International migration may also promote convergence. If the population of a low wage country is depleted by emigration—as was Ireland's during the mid-19th century—the average and marginal productivity of labour in that country is raised. From a theoretical viewpoint, the argument is general. However, from an empirical point of view, the impact on sending country wages and per capita income depends upon: (i) the ratio of emigration outflow to population size and to natural population growth (the difference between birth and death rates); (ii) the quality of the labour flowing relative to the labour staying home; and (iii) the level of remittances to the home country from individuals going abroad as a proportion of national income or exports.

Assuming labour is homogeneous, a simple elasticity measure can be used to make "back-of-the-envelope" calculations. The elasticity (e) is the ratio of the percentage change in wages relative to percent change in population. That ratio is:

$$e = ((dw/w)/(dP/P))$$

where w stands for average wages and P stands for population. Plausible estimates for this elasticity vary widely. I have seen figures ranging from -0.5 to -1.5. Let us take the most negative figure: -1.5. Surely, this figure exaggerates the impact of population decline upon the standard of living of the population exporting country. Applying these figures to crude estimates for various Asian countries given in Table 2 of Mosk (1999b) shows that the potential impact on a country's standard of living varies tremendously. For countries like the

Philippines, the impact during the 1990s appears to have been substantial. The estimates in Mosk (1999b) suggest that, as the emigrant population was around 3% of the Philippine's home populace, a dP/P figure of -3% is plausible. With an elasticity of -1.5, the implied rise in wages is +4.5%. For a nation like China, however, with a minuscule ratio of emigrants to population, it is hard to believe that the upward push in wages could be as high as +1.0%. In any case, such estimates should be taken with a grain of salt. The instrument selected for measurement is simply too crude. Moreover, assuming labour to be homogeneous is patently absurd.

Indeed, as noted in Table 9.2, if labour is heterogeneous, international migration may be a powerful force promoting divergence, for if the best and brightest emigrate—the well-educated, the talented entrepreneurs, and the ambitious—the sending country's economic leadership might be seriously wounded. Still, as Panel B of Table 2 in Mosk (1999b) shows, migrant workers' remittances calculated as a proportion of national income and national exports appear to have been substantial in the case of a number of Asian/Indian subcontinent nations. Even if the ranks of emigrants are swelled by the best and brightest, the economic impact on the sending country which operates through remittances is positive. Remittances provide sending nations with valuable foreign funds covering the costs of importing capital equipment and of licensing technology developed abroad.

A fifth factor promoting both convergence and divergence is international trade. Over the last half century the volume of international trade has expanded by leaps and bounds, especially within the Asia-Pacific region. The reasons for this expansion are manifold. Three are especially noteworthy: declining transportation costs (due to building larger and larger ships; containerization; roll-on and roll-off; the creation of Export Processing Zones; and massive investment in super ports like Rotterdam, Hong Kong, Singapore, Vancouver, and Long Beach); the forging of global agreements reducing tariffs and eliminating other institutional barriers to imports; and international convergence in incomes which has increased demand for shipment of consumer durables and capital equipment.[3] Now, in so far as promoting international trade encourages the movement of comparatively labour-intensive production to lower wage countries, it exercises a positive influence on convergence.

Moreover, as the advanced countries move through the product cycle within specific branches of manufacturing—at first, importing foreign manufactures; then, substituting domestic products for imports; moving on to become net exporters; and finally reverting to importing as the industry becomes more mature and finds it increasingly difficult to compete against lower wage competitors—industries are pushed out into other nations, especially those in the vicinity of the most advanced country in the region. As multinational corporations based in the advanced countries establish subsidiaries in lower wage nations, their activities encourage this diffusion from the core out to the periphery. For instance, in the case of Japan, both "flying geese" and "billiard ball" analogies have been used to explain the spread of labour intensive industries from Japan into the Newly Industrializing Economies (NIE) and finally into the ASEAN countries and China (Edgington and Hayter, 1999). In the "billiard ball" model, industries are pushed out into the periphery like balls. When these balls strike balls located at points in the periphery, the newly struck balls stop at the point where they collide. In the process, the collision drives balls already established at this point further out toward other countries in the periphery. For instance, textiles (and toy and bicycle production) have been largely driven out of Japan. At first, this manufacturing moved into South Korea and Taiwan. Subsequently, as television manufacturing and shipbuilding were driven out of Japan into NIEs like South Korea and Taiwan, textiles and bicycle manufacturing moved from NIEs into ASEAN countries and China.

However trade also deters convergence. One reason is the heritage of static comparative advantage developed in the past. In any economic regime, powerful interest groups—bureaucrats, business federations, and labour unions—collect economic rents from trade. When a country or region has a comparative advantage in raw materials, these rents cluster around extracting, shipping, and processing raw materials. As the traditional comparative advantage of the region slips, these groups do not ecstatically embrace the surrendering of their rents. They dig in their heels, distorting outcomes through politics.

Indeed, trade rarely operates in a political vacuum. In so far as immigration from abroad, or foreign competition in the industries which generate rents shared by business and large unions, cuts into expected future rents, campaigns to protect vested interests naturally spring up. Politicians have an incentive to respond to these campaigns, and to formulate legislation and policies which satisfy these interests. This dynamic commonly underlies trade wars, the creation of trigger price mechanisms and voluntary export restrictions which stymie or slow the diffusion of industries, with well developed niches in the advanced countries, into the less developed reaches of the globe. The Multifibre Arrangement of 1974, which protected the domestic textile industries of a number of the industrial nations, is an example of how international negotiation arising from domestic protests within these nations limited the diffusion of low wage industrial production to developing countries.

The political calculus, however, is not alone in hobbling the international diffusion of industry in response to the expansion of trade. Regional scale economies are also powerful, especially in the case of technologically intensive, and design oriented, production.[4] For instance, as Table 3 in Mosk (1999b) shows, the land vehicle (automobile, truck, and recreational vehicle) and aerospace industries are two of the world's most geographically concentrated industries. Within the industrially advanced countries of the world, some nations develop niches in one or several types of vehicles or airplanes (e.g., Germany in luxury automobiles). This specialization fosters a vigorous intra-industry trade that has been growing as a proportion of international flow during the past half century. Consumers, largely concentrated in the advanced countries, can choose from a wide menu of options, either purchasing imports or buying domestically manufactured vehicles. While this intra-industry trade promotes investment in super ports and super boats, it does not encourage diffusion of industrial production. Hence, it tends to deter convergence.[5]

Now let us consider policy. National economic policies have played an important role in convergence. The refinement of Keynesian policy and its application to fiscal and monetary policy formulation was important in the aftermath of World War II. Within the rubric of the neoclassical/Keynesian synthesis which encouraged moderate government intervention in essentially freely operating private markets, at least two major models of development emerged in the Asia-Pacific region: the "open" economy approach pioneered in North America and best exemplified by the US; and the "closed" economy approach developed in Japan and based partly upon European precedence.[6] Of course, prior to the end of the Cold War, central planning in the Communist countries provided a third model for the Asia-Pacific region. But this model has fallen into disuse since the early 1990s. It has been largely consigned to the dustbin of history.

Open economy approaches naturally emerged in frontier societies like the US and Canada where the ratio of population to natural resources and land was initially low. In such settings, agriculture always enjoyed high productivity. Hence, real wages and living standards were always generous. Migration onto the frontier stimulated an independent approach to economic activity. Distrust of financial and governmental elites was rife on the frontier. Hallmarks of the economic philosophy developed in such a setting are free entry of new firms—domestic or foreign—into markets; deregulation; immigration of foreigners and/or those born elsewhere within the nation; and financing of corporate investment through stock and bond markets in which information about corporate performance is (theoretically) available to the public.

Just as naturally, closed economy approaches sprang forth in a country like Japan where the ratio of population to resources was already—prior to industrialization—substantial. In this environment, population pushed up on land, and per capita output and the standard of living were low. Reflecting the excess supply of labour relative to resources, emigration was encouraged, and immigration was discouraged. Due to the fact that resources were relatively scarce and hence expensive, public control and long-range planning by an informed "rational" elite offered the best chance for sustained development through the import of technology. Free entry into new markets was not embraced by bureaucrats in governments or in giant corporations who viewed their visions as more objective and dispassionate than that of the general populace.

In the closed model, elites attempt to control, limit, and direct the flow of information. For instance, in Japan, regulation of financial markets through a tightly controlled banking sector which was the principal source of private corporate investment funding during the period of catching up with the West, gave politicians and bureaucrats leverage over the pace and characteristics of private sector economic activity. Easy entry of foreign firms was disdained as foreign corporations could not be as easily regulated as domestic enterprises. Barriers were erected to the import of foreign goods and foreign investment. Finally, the Japanese government focused on developing human capital through the creation of a technologically oriented education system so that dependence on foreign capital was minimized, and native experts could be substituted for foreign counterparts.

The distinction between open and closed economies is highly stylized and should not be taken as a literal description of development in either North America or in Japan, nor should the role of ecology and population density be exaggerated. For instance, British control over significant elements of Canada's agenda prior to World War I, and the presence of numerous British and Scottish labour union organizers in frontier areas like British Columbia, encouraged the diffusion of a government planning/regulation approach to development in British Columbia which shares some features with the closed approach pioneered in Japan. Nevertheless, the distinction points to the emergence of two philosophically distinct models of economic advance which the emerging countries of the Asia-Pacific have emulated.

Finally, a third type of policy has encouraged convergence: environmental concerns as captured in national and international regulation. As a clean environment is usually viewed as a luxury within domestic politics, demand for environmental quality typically rises with income per capita, as does demand for luxury goods like jewellery. Hence, low income countries enjoy a competitive edge in pollution-causing industries like iron and steel manufacturing. Their governments find themselves contending with less vocal environmentalist movements than do those in advanced countries. An excellent example of how catching up with advanced countries spurs the creation of an environmentalist movement is illustrated by Japan's ignoring of pollution concerns before the 1970s, followed by its growing embrace of regulation over environmental degradation (Ramseyer, 1996, Chapter 3).

Finally, we need to consider a seventh factor listed in Table 9.2: the growth of mega-cities in the third world. As is shown in United Nations (1995), most of the largest metropolitan centres in the near future will be in the developing world. During the 19th century, most international migration was from rural areas to rural areas. Today, most migration involves individuals moving from large cities to large cities. Thus, the proliferation of mega-cities in the developing world tends to promote emigration from these countries. The mega-cities also harbour scale economies which promote modernization of infrastructure and the attendant diffusion of technology. Thus - depending on the nature of emigration - mega-city growth promotes convergence. But, as noted in Chart 1, mega-city growth in the developing regions of the Asia-Pacific joins the six factors previously considered, in promoting both convergence and divergence.

In sum, at a theoretical level, it is difficult to predict whether divergence or convergence will occur, either on a global scale or within the Asia-Pacific region. The outcome depends on the strength of the opposing forces. Hence the issue is empirical. And, as already noted, it appears that convergence is winning out over divergence within the Asia-Pacific. Thus, it is reasonable to conclude that the forces promoting convergence have been stronger than those engendering divergence over the course of the last half century.

Post-Cold War Expansion in the Two Main Engines of Asia-Pacific Growth: The United States and Japan

With the collapse of the central planning/non-market oriented model of economic development, three major paradigms compete within the Asia-Pacific region: the North American model with its emphasis on open markets; the semi-closed/semi-open Western European or Third Way model with its stress welfare and redistribution; and the closed model best exemplified by Japan. Of these three models, only two are really relevant in the Asia-Pacific context: the open and closed models. The reasons for this are geographic. The two dominant economies of the region are the American and Japanese economies; and the climate of the Asian region—namely, the monsoon rain pattern which encourages dependence on rice cultivation—promotes high population densities in agriculture. This is because per hectare calorie output is high under rice cultivation, and hence farmland can be subdivided and subdivided until large numbers of agriculturalists are sustained on limited land areas.

In short, the Japanese model of development has special appeal in the most densely populated sub-regions of the Asia-Pacific. However, it vies with the North American approach which is harnessed to the mighty US economic engine, and recent economic trends over the last decade seem to suggest that the American model is performing better than the Japanese model, or at least the American economy is currently showing better growth prospects than is the Japanese economy.

As the following figures on mean annual growth rates (%) in real GDP over the 1992-96 period show, growth in the US has been remarkably steady and strong, while growth in Japan, Europe, Canada, and Pacific Canada has been jerkier and/or less vigorous:[7]

	Country/Province				
Item	Japan	US	Europe	Canada	BC
Range of Growth	3.6=3.9-0.3	1.5=3.5-2.0	3.2=2.9-(-.3)	3.8=4.7-0.9	2.9=4.0-1.1
Mean Annual Growth	1.46	2.56	1.56	2.34	2.44

To be sure, the period covered is short. It would be unwise to conclude that US-style growth is superior to Japanese-style growth: growth may be leapfrogging around the world. It may pick up in one region and run its course there, only to move on to other global economic hot spots. However, the data may be telling us something very important about the long-run viability of the closed or semi-closed models of growth. For instance, are we to conclude that Japan is going through a "climacteric" at the end of the 20th century, similar to the climacteric suffered by the United Kingdom at the close of the 19th century?

The debate rages about Japan's slowdown during the 1990s. Some commentators take a short-run view, arguing that the lethargic pace of growth is due to the bursting of the bubble economy. And they view the bubble phase itself—during which asset prices for land and stocks soared—as an aberration fuelled by a combination of misguided domestic monetary policies, the globalization of financial markets, and speculation.

However, other commentators take a long-term approach, pointing to basic structural shifts in the Japanese economy and population which—in my view, because of path dependence (hanging onto behavioural patterns inherited from the past)—have not been sufficiently thoroughgoing. As a result, Japanese growth has been highly uneven and erratic since the early 1970s. During the 1980s, growth was too rapid; and, after the early 1990s, growth has been too slow. In any case, the inference is that, unless it makes vigorous adjustments, Japan may have indeed entered a "climacteric" period of relative decline.[8]

According to the long-term view, several aspects of Japan's closed economy approach to development are seen as causing the bubble economy and its subsequent collapse. For instance, even before Japan's malaise in the 1990s became noticeable, Lincoln (1988, 1990) pointed out that the surplus of savings over investment posed a major problem for Japan. Indeed, the excess of savings over investment is the mirror image of Japan's dependence on exports for generating growth in the post-1975 period. During the era of High Speed Growth (1955-70), when Japan caught up with the advanced industrial countries, rising savings was a necessary concomitant of investment-led growth fuelled by high expected rates of return on new capital construction, which in turn depended upon substantial gains in total factor productivity stemming from importing and adapting foreign technology, for the Japanese government pursued a closed economic strategy. It made it difficult for foreign companies to enter the Japanese market, and it actively discouraged foreign financing of domestic investment. Rather, the Ministry of Finance and the Bank of Japan encouraged the growth of a financial system whereby Big City Banks channelled funds into long-term investment projects of major enterprises. The banks relied upon implicit guarantees of bailout by the central bank to effectively insure the financial intermediaries.

As long as rates of return on long-term investment were high—gigantic rates of total factor productivity growth ensured they would be high—the banking system thrived in the closed economy environment. Massive bailout was not required. By the early 1970s, however, Japan had reached the technological frontier in most industries. Thus, rates of total factor productivity growth plummeted, bringing down rates of return on new capital acquisition. The incentive to invest in new plant and equipment was eroded, and overall private investment fell. Thus, Japan entered a period when savings exceeded investment. An equivalent statement is that Japan began to run huge trade surpluses; it began to export capital, and the yen marched upward on international exchange markets. Higher values for the yen relative to the US dollar (which served as the world's main clearing currency in handling international transactions) drove up Japanese labour costs relative to those in other countries. As a result, Japanese investment flowed into lower wage countries, especially into those Asian countries nearby Japan, but Japanese capital also flowed into North America and Europe. These latter flows were driven in part by the desire of Japanese companies to stave off domestic content legislation and the impact of voluntary export restraints imposed by its advanced industrial trading partners.

As Japanese domestic investment fell and Japanese companies increasingly focused on acquiring capital in other regions of the globe, Japanese banks faced a conundrum. During the heyday of the closed economy they had concentrated on lending to corporations. Now they faced declining demand for these loans on the part of major corporations. Still awash in savings deposits, they turned to riskier corporate loans and to real estate, setting the stage for a speculative bubble in asset prices, and for the subsequent collapse of the bubble. This story, which emphasizes excess savings and a financial system that failed to make the transition out of the era of high speed growth and Miracle Growth catch-up with the West, is told by Lincoln (1998) and by Sato (1999).

In my view, the long-term failure to adjust to its convergence with the advanced industrial world offers the best explanation for Japan's current lacklustre performance. In effect, to adequately adjust to convergence, Japan should have completely opened up during the early 1970s. It should have abandoned its closed economy approach, but path dependence is very strong. Hence, during the 1980s Japanese policy makers operated under a bizarre illusion that high speed growth would return to Japan, and behaved accordingly. However, the days of 5% per annum rates of total productivity growth ended right before or during the early 1970s (Nakajima, Nakamura, and Yoshioka, 1999). Growth during the 1980s was too strong, and as a result, Japanese growth during the 1990s has been too slow. To be sure, Japan will return to a path of moderate growth, at least for the medium-term, but Miracle Growth is over. Whether bureaucrats and politicians heady with the success of the closed economy of the 1950s and 1960s accept this fact is another matter.

Japan has been adjusting. Japan has been opening up and it has been abandoning its closed economy institutions. However, the pace at which it does so is painfully slow. For instance, Japan has become a leading defender of free and open trade. It has (somewhat reluctantly) joined the other advanced countries in developing policies for stopping environmental degradation, and it has even (slowly, and by a mere crack) opened its doors to immigrants from Latin America and from Asia (Meissner, Hormats, Walker, and Ogata, 1993). In so far as it has opened up, it has converged toward a model of economic behaviour which is not exclusively rooted in a closed economy model, and as Japanese industries—especially raw materials heavy industry—have been moving into other regions of Asia and out of domestic production, Japan's voracious appetite for some raw materials has slackened off.[9] Japan's character as a country exporting manufactures and importing raw materials is changing. Its import of manufactures is rising, and its import of raw materials (as a share of imports) is falling off.

In short, Japan is adjusting to convergence in the Asia-Pacific region. It is adjusting to its own convergence with the most advanced economies of the region, and it is also adjusting to convergence of other countries in Asia to levels of per capita income approximating its own. Perhaps the slowness of Japan's opening up is due in part to the fact that it has taken several decades since 1970 for the NIEs and some of the ASEAN countries to make substantial progress on the road to convergence with Japan; but path dependence has also slowed Japan.

Now let us consider British Columbia as a case of adjustment of a wholly different sort to convergence within the Asia-Pacific region. As we have seen, British Columbia's recent performance has been erratic. By the standard set by the US, British Columbia's performance has been remarkably unstable.

If Japan has been adjusting by opening up, and by importing more manufactures (that is, by changing its economic structure), British Columbia has been adjusting by diversifying out of its traditional raw materials industries. However, if path dependence has slowed Japan's convergence toward a more balanced economy, increasingly open in orientation, path dependence has also deterred British Columbia from rapidly adjusting to the globalization of raw materials extraction, processing, and shipping which has been ongoing for the last several decades. Convergence has been undermining Japan's closed economy approach. It has also undermined British Columbia's reliance on raw materials, for Pacific Canada is now facing a whole new regime in terms of trade between raw materials and manufactures. In this new regime, raw material prices have systematically dropped relative to the price of manufactures; but bureaucrats and enterprises that once extracted rents out of the British Columbia forestry, mining, and fishing sectors have a vested interest in fighting to keep the Pacific Canadian economy oriented around these declining sectors in which it once possessed a strong comparative advantage. So, as in Japan, path dependence is slowing the pace of adjustment.

Nevertheless, adjustment has been taking place in British Columbia. The province is adjusting to the growing importance of demand for manufactures and technologically sophisticated output in the Asia-Pacific region, especially the western US and Japan. The expansion of greater Vancouver as a Pacific-oriented trade port is one result (Davis and Hutton, 1989; Hutton, 1997; and Resnick, 1985). As Vancouver's status as an Asia-Pacific oriented metropolis has been strengthened, the interdependence of the greater Vancouver region with the resource-oriented remainder of the province has shrunk.

It is important to recognize that these recent economic adjustments in Japan and British Columbia to Asia-Pacific convergence have been taking place during a period of uneven performance within the NIEs and ASEAN sub-regions of the Asia-Pacific. This unevenness has many different sources, and the causes within any one country are not necessarily identical to those in other countries of the region. At a general regional level, however, the jerkiness of recent Asian performance stems from the growing importance of the American model of investment—in which commitments are mainly put on a short-term basis—within the Asia-Pacific. Indeed, Rodrik (1999) argues that short-term capital investment, which moves around quickly from one place to the next, is growing within Asia, and it is generating instability there. Speculative booms and busts are growing in frequency and in magnitude.

Still, the main thrust of this section is that longer term trends toward Asia-Pacific convergence dominate the process of adjustment within the economies of Japan and British Columbia. Convergence is not simply achieved through trade, capital movements, technology transfer, and the adoption of policy models, however. It is also conditioned by the migration of persons. Let us see how migration has shaped the process of adjustment in Japan and in British Columbia.

Demographic Integration with the Rest of the Asia-Pacific: Pacific Canada versus Japan

Japan and Canada are at opposite poles as far as migration is concerned. Canada, and especially Western Canada, has been relatively open to immigrants for centuries. To be sure, Canadian openness has hardly been unconditional. For instance, until recently, Canada was not as open to Asian immigrants as it was to immigrants from Europe and the US. By comparison, Japan has been almost completely closed for centuries. After it "opened up" to the West in the 1850s and 1860s, it began to encourage emigration. However, immigration has been tightly controlled, and, until recently, it was largely restricted to Koreans and Chinese brought in involuntarily during the late 1930s and during the World War II period. Many of the descendants of these immigrants are still classified as foreigners residing in Japan; that is, as immigrants.

Path dependence remains strong in the Japanese case. As Table 5 of Mosk (1999b) shows, even after Japan began to selectively open its labour market to immigrants, the impact of immigrants on Japanese labour market activity is minimal. The price of keeping labour markets relatively closed is substantial. Taken in conjunction with a low birth rate, Japan's closed demographic policy can potentially spawn demographic decline, and in the international competition to secure highly skilled, talented professionals or wealthy entrepreneurs from other parts of Asia, Japan is losing out to the relatively open countries of settlement in the Asia-Pacific; that is, to the US, Canada, and Australia.

Path dependence is far less strong in the Canadian case. It exists in the sense that the annual magnitude of immigrants allowed into Canada has not varied much over the postwar years. While annual fluctuations in numbers admitted has occurred—adjusting to changes in federal immigration policy driven in part by business conditions—the overall volume is quite stable.

Still, switching from a focus on absolute volumes to a focus on the ethnic origin of immigrants reveals dramatic changes in Canada's immigration. As is seen in Panels A and D of Table 2 in Mosk (1999b), Canada is similar to the two other settlement countries of the Asia-Pacific region—the US and Australia—in concentrating on absorbing persons of Asian descent over the last several decades. This focus on absorbing Asians is relatively recent. To some extent, it reflects convergence: all three countries have introduced business class immigration classifications, which encourages the wealthy of Asia to seek settlement across the Pacific. For example, this accounts for the high percentage of persons of Hong Kong origin in Canada's recent immigration inflow (Panel D of Table 5 in Mosk, 1999b). It also reflects a changing international political dynamic, as pressure is applied on all three settlement countries by the lower income nations in the Asia-Pacific region. For instance, in all three countries immigration during the 19th and early 20th centuries mainly involved absorbing Europeans. In the case of Canada and Australia, the switch away from bringing in Europeans has principally meant welcoming persons from the Asia-Pacific region. In the case of the US, the immigrant pool has been increasingly populated by persons from Central and South America, and from Asia. In short, the discontinuity in Canada's immigration is part of a general pattern of reorienting immigration in the settlement countries of the Asia-Pacific region. That reorientation has increased demographic integration across the Pacific.

In sum, demographic integration is increasing in the Asia-Pacific region. It is increasing because the traditional countries of settlement—the US, Canada, and Australia—are absorbing increasingly large numbers of Asian immigrants; because Japan is gradually opening itself up to immigration; and because a number of the NIEs and ASEAN countries have encountered labour shortages in some sectors, and have opened themselves up in a limited way to immigration from other parts of Asia.

In short, demographic integration is responding to convergence in the Asia-Pacific region. As former lower income countries in the region close the gap between themselves and the advanced nations through high rates of economic growth, they switch from being net emigrant sending to net immigrant receiving states. As they make this transition, migrants from lower income nations in the region join the labour forces of the advanced nations to fill the excess demand for labour created by the rapid rates of growth in these countries. Thus, labour migration expands integration; and it is increasing due to political pressure applied to the countries of settlement and to perceived economic self-interest in these nations stemming from convergence. Thus, convergence begets demographic integration which, in turn, contributes to convergence.

Bilateral Economic Integration and Regional Convergence: Pacific Canada and Japan

So convergence does seem to be spurring on economic and demographic integration in the Asia-Pacific region. However, because of path dependence and the difficulties of jump-starting development in some sub-regions, integration is taking place slowly. Moreover, it is occurring at the level of the entire Asia-Pacific region. As a consequence, we cannot say whether any two countries—or regions of countries—are becoming more or less integrated on a bilateral basis. Indeed, one could argue that the declining demand for raw materials in Japan —resulting from raw material intensive industries moving out of Japan into the NIEs, the ASEAN countries, and China—can diminish bilateral integration between Japan and its regional raw materials suppliers. Pacific Canada is a source of coal, minerals, fish, and timber for Japan. While the US absorbs the largest share of British Columbia's exports in general, Japan is British Columbia's biggest demander of some raw materials exports like coal. For this reason, while convergence could strengthen British Columbia's export flows to other countries in the Asia-Pacific region with the exception of Japan (but including the US), it could undermine bilateral trade between Japan and British Columbia.

In particular, the paradox arising from the divergence between long-run and short-run implications of convergence in the Asia-Pacific could incite Canadian exporters in the Pacific to draw incorrect conclusions about the future of economic integration between Pacific Canada and Asia. Japan is in a period of slow growth, and even when the growth impulse returns, the long-run secular effect of convergence is working to reduce Japan's need for traditional British Columbia exports. Thus, from a bilateral viewpoint, Japan will probably diminish as a demander of Pacific Canada's traditional exports, regardless of the economic growth rates in Japan itself. However, in the long run, convergence should bolster British Columbia's exports.[10] For this reason, investors in British Columbia should not draw unwarranted conclusions about the future of regional integration from the bilateral experience of British Columbia with its largest Asian trade partner, Japan. The paradox is that it is natural to see bilateral experience with Japan as a harbinger for diminishing interaction with Asia when, in fact, it is really a harbinger for a brighter regime of economic integration in the future.

Recent statistical trends in British Columbia-Japan trade support a dismal scenario for economic integration between Japan and British Columbia (Table 6 in Mosk, 1999b). British Columbia's major raw materials commodity exports to Japan have tended to drop precipitously, with the exception of newsprint. These declines have taken place at a time when—in terms of Canadian dollars —Japanese income has risen markedly. In short, during a period when Japan's capacity to purchase raw materials from British Columbia has dramatically improved, its actual purchases have diminished.

The one bright side would seem to be tourism. According to Table 6 in Mosk (1999b), it is apparent that tourism responds positively to Japan's purchasing capacity, but the importance of Japanese tourism in British Columbia should not be exaggerated. In comparison to the volume of tourists coming from other Canadian provinces, and those coming from the US, the number of Japanese is not large.

So it would seem that economic integration between British Columbia and Japan is weakening; but this is not necessarily so. As Japan has adjusted to convergence, as it has increased its imports of manufactures relative to raw materials, and as it has stepped up its import of tourist services, the nature of British Columbia's Japan oriented business base has shifted. Consider the figures given in Table 9.1 concerning Canadian companies doing business with Japan. As can be seen, the percentage of Canadian companies doing business in Japan (and/or with Japanese organizations) has increased at an especially rapid rate in British Columbia during the 1990s. Indeed, the absolute increase in the number of companies located in British Columbia is almost 40%. A large number of these companies are in manufacturing, construction/building equipment and supplies, and trade related services. The increases have not been in the traditional raw materials industries.

In short, as Japan's economy has adjusted to convergence in East Asia, it has engendered structural change in British Columbia through its impact on the demand for goods and services in Pacific Canada. Japan's adjustment has encouraged diversification in Pacific Canada. A similar story can be told about British Columbia exports to the US. In 1990, 9.2% of British Columbia exports to the US were in the machinery and equipment sub-sector of manufacturing. In 1998 this had risen to 15.6%. Including the west coast of the US as a key part of the Asia-Pacific region, we can say diversification in British Columbia is being driven partly—perhaps mainly—by convergence in the Asia-Pacific region.

Conclusions

Over the last half century, the Asia-Pacific region has been increasingly integrated in both the purely economic and purely demographic dimensions. Key to this trend toward integration is economic convergence in the region.

A number of forces—capital accumulation, education, diffusion of technology, migration, international trade, policies, and mega-city growth in developing Asia—have been at work in generating convergence, and hence, integration. From a theoretical point of view these seven forces can promote both convergence and its opposite, divergence. Thus, the question why the trend toward convergence and integration has occurred is largely an empirical issue. Theory alone fails to explain why it has taken place. In the overall opposition of convergence and divergence, impacts promoting convergence have dominated over impacts promoting divergence.

Both Japan and Pacific Canada have been adjusting to convergence. Japan has been opening up, exporting capital and technology, and importing a growing number of manufactures. Increasingly, Japan is engaging in intra-industry trade; and British Columbia's economy is undergoing diversification. These adjustments taking place on opposing sides of the Pacific have affected the bilateral interaction between Japan and British Columbia.

At present it seems that British Columbia's stake in Asia is in decline. The US has grown as a trade partner, and Japan—British Columbia's largest Asian trade partner—has been falling off as a demander of Pacific Canada's exports. In the long run, however, convergence in the Asian sub-region of the Asia-Pacific will re-emerge as an engine of growth in British Columbia. The ongoing secular trend toward convergence which has driven integration across the Asia-Pacific will see to that.

Endnotes

1. This is a revised version of the paper "Structural Change in Japan: Its Impact on Asia and on the Pacific Rim of North America" that I prepared for the University of Victoria/ National Sun Yat-Sen University (Taiwan) Symposium at Dunsmuir Lodge in August 1999. I am grateful to the discussant Alan Hedley, and to the conference participants, for helpful comments. Due to space limitations I cannot present all of the tables which I prepared for the presentation. These tables appear in Mosk (1999b), which can be accessed on the internet at http://riim.metropolis.net/research-policy/research-policy2/ papersel.html #1999.

2. For the estimates upon which this paragraph rests, see Mosk (1999b): Table 1.

3. On the buildup of the super port of Vancouver, see Vancouver Port Corporation (1990). On the importance of seaports and of infrastructure for the industrialization of Japan after 1870, see Mosk (1999a). On the development of Export Processing Zones, see Shoesmith (1986).

4. For the importance of regional scale economies, and the linkage of these scale economies to the regional concentration of pools of skilled labour, see Mosk (1999a).

5. To some extent, diffusion has taken place in the Asia-Pacific region outside of Japan and North America. For instance, in South Korea during the 1970s and 1980s, shipbuilding, automobile manufacturing, and consumer electronics became well established branches of industry. During the late 1980s and 1990s, subcontracting (and the manufacture of components) for Japanese corporations has grown throughout Asia, especially in the ASEAN countries; and, as shown by Hayashi (1999), with establishment of subsidiaries of Japanese companies in the ASEAN countries, technology has been transferred from Japanese industrial laboratories to their subsidiaries. Hayashi (1999) notes that the length of time required for ASEAN to initiate production of technologically sophisticated products has vastly shortened over time. Lags between the inception of new product manufacturing within Japan, and the beginning of similar production in foreign subsidiaries, have been dramatically squeezed down, falling to a year or less in some cases.

   The situation on the west coast of the US and the west coast of Canada is quite different. For instance, vehicle production is well developed in California (which enjoys a population base comparable to Canada's), and Boeing in Seattle is a leading aerospace manufacturing enterprise. With its smaller domestic market, manufacturing within Canada historically has been concentrated in one region, namely in the Ontario-Quebec zone near the St. Lawrence seaway. However, this state of affairs is in flux, for as the discussion in Section V of this chapter reveals, the economic structure of Pacific Canada is becoming more diversified, and this diversification is being driven by the expansion of manufacturing activity in British Columbia.

[6] The general distinction between "open" and "closed" economies is originally due to Simon Kuznets.

[7] I am grateful to Alan Hedley for reworking figures given in Table 4 of Mosk (1999b) which yielded the numbers given here. For sources for the data see the notes to Table 4 in Mosk (1999b).

[8] A recent front page article in the New York Times presents a case for a "Japanese climacteric" (see Kristof, 1999). Kristof emphasizes the excessive building of infrastructure and population decline in contemporary Japan, and he ties the theme of population decline to the unwillingness of Japan to more freely open up its borders to immigration. That is, he argues that Japan remains too closed, especially on the demographic side. This is a position with which I concur.

[9] As early as the beginning of the 1970s, Japanese policy makers were aware of a need to downsize raw material intensive, polluting industries (Tanaka, 1972). For a discussion of structural shifts within sub-sectors of Japanese manufacturing during the 1970s and early 1980s, see Dore (1986). The decline in Japan's manufacturing base has led to degradation of the status of the Ministry of International Trade and Industry (MITI) within the tiny club of powerful ministries in Japan. It has also led to a reorientation of MITI's focus toward encouraging imports and controlling pollution. For details, see Kohno (1999).

[10] It is important to note that Japan is the only country in Asia that uses large quantities of British Columbia's timber/processed log exports, because Japan alone makes heavy use of wood in housing construction. Moreover, it should be noted that due to technological changes in the pulp and paper industry, British Columbia's traditional position in this industry is being eroded by competition from Asia, Russian Siberia, and Europe.

References

Canada-Japan Trade Council (Undated a). *Companies in Canada doing business with Japan, 1994-95.* Ottawa: Canada-Japan Trade Council.

Canada-Japan Trade Council (Undated b). *Companies in Canada doing business with Japan, 1997-98.* Ottawa: Canada-Japan Trade Council.

Davis, H. C., and Hutton, T. A. (1989). The two economies of British Columbia. *BC Studies.* 82, 3-15.

Dore, R. (1986). *Flexible rigidities: Industrial policy and structural adjustment in the Japanese economy, 1970-80.* Stanford: Stanford University Press.

Edgington, D. and Hayter, R. (1999). Japanese electronics firms in Asia-Pacific: Examining the flying geese model. Paper presented to the UBC Interdisciplinary Conference, *Japanese Business and Economic System: History and Prospects for the 21st Century*, February 1999.

Goto, J. (1990). *Labor in international trade theory: A new perspective on Japanese-American issues.* Baltimore: The Johns Hopkins University Press.

Hayashi, T. (1999). Technology transfer in Asia in transition. Paper presented to the UBC Interdisciplinary Conference, *Japanese Business and Economic System: History and Prospects for the 21st Century*, February 1999.

Hutton, T. (1997). The Innisian core-periphery revisited: Vancouver's changing relationship with British Columbia's staple economy. *BC Studies*, 113, 69-100.

Japan Institute of Labour (Various dates). *Japan labour bulletin.* Tokyo. Japan Institute of Labour.

Kohno, M. (1999). The changing role of MITI and its implications for US-Japan relations. Paper presented to the UBC Interdisciplinary Conference, *Japanese Business and Economic System: History and Prospects for the 21st Century*, February 1999.

Kristof, N. (1999). Empty isles are signs Japan's sun might dim. *The New York Times*, August 1, 1999.

Lincoln, J. (1988). *Japan: Facing economic maturity.* Washington, DC: The Brookings Institution.

Lincoln, J. (1990). *Japan's unequal trade.* Washington, D. C. The Brookings Institution.

Lincoln, J. (1998). Japan's economic mess. Paper presented to the *Japan Economic Seminar*, April, 25, 1998.

Meissner, D., Hormats, R., Walker, A. G., and Ogata, S. (1993). *International migration challenges in a new era.* New York: The Trilateral Commission.

Mosk, C. (1999a). *Technology, economy, city: Osaka and the Japanese industrial belt.* Book manuscript under review.

Mosk, C. (1999b). *Convergence and divergence in the Asia-Pacific: Economic and demographic integration between Asia and Pacific Canada.* Research on Immigration and Integration in the Metropolis (RIIM), Working Paper 99-21.

Nakajima, T., Nakamura, M., and Yoshioka, K. (1999). Japan's economic growth: Past and present. A paper presented to the UBC Interdisciplinary Conference, *Japanese Business and Economic System: History and Prospects for the 21st Century*, February 1999.

Nakamura, T. (1995). *The postwar Japanese economy: Its development and structure.* Tokyo: University of Tokyo Press.

Ramseyer, J. M. (1996). *Odd markets in Japanese history: Law and economic growth.* Cambridge: Cambridge University Press.

Resnick, P. (1985). BC capitalism and the empire of the Pacific. *BC Studies*, 67, 29-46.

Rodrik, D. (1999). *East Asian mysteries: Past and present. NRER Reporter*, Spring, 7-11.

Sato, K. (1999). Japanese economy in the 1990s: What has gone wrong? The consequences of excessive financial deepening. Paper presented to the *Japan Economic Seminar*, April, 1999.

Shoesmith, D. (Ed.) (1986). *Export processing zones in five countries: The economic and human consequences?*, Asia Partnership for Human Development.

Stalker, P. (1994). *The work of strangers: A survey of international labour migration.* Geneva: International Labour Office.

Tanaka, K. (1972). *Building a new Japan: A plan for remodeling the Japanese archipelago.* Tokyo. The Simul Press.

United Nations, Department of Economic and Social Affairs, Information and Policy Analysis, Population Division (1995). *World urbanization prospects: The 1994 revision.* New York. United Nations.

Vancouver Port Corporation (1990). *Port Vancouver.* Vancouver. Vancouver Port Corporation.

Plate 12 A sculptured stone lion guarding a Chinese temple ➧

佑群黎

10

Optimal Adjustments of Taipei's Trade Policy toward Beijing after the 1996 Missile Crisis

Tru-Gin Liu

Associate Professor, Institute of Economics, National Sun Yat-Sen University

Introduction

The objective of this chapter is to study the optimal adjustments of Taipei's trade policy toward Beijing within the framework of triangular interactions among Beijing, Taipei, and Washington. The mode of such interactions has evolved into a new era over the last decade. Needless to say, this framework has been intertwined with conflicts of cooperation and competition over various issues. It has been emerging as very complicated when the Taiwan issue is involved. On the basis of this thinking, this chapter argues that Taiwan authorities will need to consider their trade policy, including its investment policy with China, in a broader framework involving the changing pattern of interactions among Beijing, Taipei, and Washington.

The players involved in the interaction between Beijing and Taipei are not as straightforward as they appear. In addition to the two parties across the Taiwan Straits, there are other latent players involved. For example, Klintworth (1995) has pointed out that, from the viewpoint of the US and Japan, the fact of Taiwan's increasing dependence of trade on China will help China to develop as the next superpower of the world. Thus, for the US and Japan, it becomes more important strategically to prevent Taiwan from being influenced by China. Klintworth's argument reflects the fact that trade across the Straits is not only a matter involving China and Taiwan, but also the US and Japan. Putting it simply, we can consider trade across the Straits as a sub-game, which is part of a larger game involving other latent players. Consequently, a more comprehensive analysis of Taiwan's trade policy toward China will have to take into account the mode of interaction between the US and China, if we think of the US as the leader of a group comprising countries such as Japan and Australia.

Lin and Lo (1998) analyse trade issues across the Straits from the perspective of game theory, but fail to recognize a few assumptions implied in their model. For example, they assume that the two parties involved are denied the possibility to communicate with each other. They also fail to consider whether there are any devices normally used for making mutually advantageous deals, and whether the threat of China's unusual practice is credible or not. As a result, their model does not fit reality very well. Taking the 1996 missile crisis as an example, although the role of the US as an arbitrator is somewhat obscure, the fact that the US was involved during certain stages of the crisis does suggest that there is at least one other player in addition to Beijing and Taipei. It is highly questionable to assume that Taiwan and China are both isolated from the outside world, and are not in communication through an arbitrator.

With regard to the so-called hostage issue (China holding Taiwan hostage), does it exist in reality? During the 1996 missile crisis, the likelihood that China and the US might have gone to war cannot be ruled out as the US carrier Independence sailed into the Taiwan Straits. Under these circumstances, the attempt to avoid an outbreak of military action was probably not the sole concern of the US. Among other concerns, the US authorities would have had to consider the price if the US were involved. Therefore, as far as the Taiwan issue is concerned, the US would have had to consider how to balance its political interest against its economic interest.[1]

In a dynamic game, we normally assume that the rationality of players is common knowledge. Players select their best strategies or actions in their own interest. However, during the 1996 missile crisis, the dangerous actions of Beijing can hardly be seen as "rational" if we consider the outbreak of China-Taiwan warfare as a zero-sum game or even a negative-sum game. To explain Beijing's extreme reaction, I think the picture would be clearer if we elucidate the hostage issue by considering the missile crisis as one of Beijing's strategies against the US.

If we view the above issue from this perspective, we have the hypothesis that China's intention is very likely to take Taiwan as a hostage, and thus force the US government to make more concrete official declarations about Taiwan's international status, which must be acceptable to both China and the US. The purpose of Beijing's strategy is to change the sequences of the game among the three main players. Before the missile crisis, we can consider the US as an initial mover in the first stage of the game, while Beijing and Taipei select individually their best strategies in the second stage. Or, we can consider that, in the first stage, the US and Taipei select their best strategy based on their alliance relationship, while in the second stage Beijing selects its own optimal strategy. However, since the missile crisis, the structure of the game has been changed. Beijing and the US negotiate directly over the Taiwan issue first, and afterward Taiwan has to select its best response in light of the result of the negotiation. It is likely that this new structure of the triangular game will have multiple stages.

After Clinton's visit to China in 1998 and the unprecedented US presidential announcement on one-China policy, it seems obvious that the ongoing structure of the game has been characterized by negotiation between the US and China. Let us consider the following scenario: Beijing and Washington are bargaining over the Taiwan issue. The target under negotiation is the extent to which the US supports Taiwan politically.[2] Suppose that the extent of the support can be standardized by an index. We can then think of an optimal index, given any reaction by Taiwan, which can maximize the joint interest of the US and China. After that, Taiwan makes its move accordingly. However, Taiwan's reaction may in turn disturb the initial agreement of the US and China, and force them to negotiate again. This process will not end until the equilibrium is reached.

Assume that Taiwan's objective is to maximize its own interest. In the face of the new structure of the triangular game, one would then wonder whether or not it is optimal for Taiwan to maintain its current trade policy toward China, that is, the policy of refraining from being impatient to invest in China. Or, to put it straightforwardly, what should Taiwan do to capitalize on fundamental conflicts between Washington and Beijing to serve Taiwan's own interest?

The arrangement of the chapter is as follows: The next section describes changes in the Asian-Pacific region, and the changing role of Taiwan over the past two decades. The third section builds the analytical model. It focuses on the exchange of political interest and economic interest between China and the US, as well as Taiwan's optimal adjustments in its trade policy toward China in response to such an exchange. Section four concludes the chapter.

Changes in the Asian-Pacific Region and the Changing Role of Taiwan

Establishing commercial links with China during 1986-87 was of substantial importance for Taiwan. Over the last two decades, most countries avoided making official contacts with Taiwan, either out of respect for China's diplomatic policy or for practical reasons. Ironically, it turned out that the two Chinas began to make contact with one another while each side maintained its own one-China policy. The curiosity for the West is why China and Taiwan maintain their respective notions of the legitimacy of China, and hope that other countries will accept the one-China policy claimed by each side. While other countries have made up their minds to make a choice between the two "one-China" policies, it seems odd to them that contact between China and Taiwan is becoming more and more intimate, covering many areas such as trade, commerce, investment, and cooperation in technology, law, business, and transportation.

Without a doubt, the role that Taiwan plays in East Asia is contingent upon the dynamics of the political and economic environment in the Asia-Pacific area. If Taiwan were to develop its economic and trade ties with a superpower while under the rule of the other, as it was before the disconnection of its diplomatic links with the US in 1979, then such a development would be constrained by the country dominating Taiwan. However, the historical background of the period when Taiwan began to establish commercial links with China in 1986 is unique. By the 1980s, the mode of interaction among the superpowers surrounding Taiwan was no longer similar to a game of entry deterrence. Instead, it had become a game of multilateral communication and strategic coordination. In such circumstances, the big swing in Taiwan's trade policy toward China has pushed Taiwan into the core position among the three major powers, that is, China, Japan, and the US. Therefore, it can be said that Taiwan's independent existence is dependent upon the interactive balance among the three superpowers.

Taiwan's geographic position in the Asia-Pacific area, right on the cross between West Pacific, South East Asia, and South China Sea, is distinct. Any superpower surrounding Taiwan certainly wishes to control it. Before 1945, Taiwan had been under the control of a powerful nation because of the asymmetric relations of the three superpowers in the Asia Pacific region. The game played by these powers was very similar to that of entry deterrence with Japan as the incumbent and China and the US as potential entrants. However, the situation has changed along with the emergence of the following three factors: the defeat of Japan, the retreat of the US army from Vietnam, and China's economic reforms starting from 1979. Asymmetric relationships in the Asia Pacific region among the three superpowers have gradually moved toward symmetric and well-balanced relations. Since exclusive power over Taiwan is no longer possible, hostility will not work. Instead, cooperation-cum-competition has emerged as the mode of interaction among China, Japan, and the US, and is largely based on coordination mechanisms.

Taiwan has been cautious not to utilize aggressively the change in interaction mode among the big powers surrounding her before 1986. The triangular relationship confronting Taiwan in the 1980s was biased toward the circle linking the US, Japan, and Taiwan, but not that linking the US, Japan, and China. Furthermore, owing to the hostility across the Straits, Taiwan had to avoid being in contact with China, but had to watch the US and Japan making deals with China. Therefore, the weight of the triangular relationship was biased toward the US and Japan. By 1986 and 1987, there was an emerging force pushing behind Taiwan, and eventually leading Taiwan to be the centre of a new triangular relationship.

As a matter of fact, this force originated from economic factors. In 1986 and 1987, Taiwan's currency experienced high appreciation, which increased the costs of Taiwan's manufacturing sector. To cope with this situation, a number of Taiwanese firms looked for investment opportunities overseas to maintain the competitiveness of their products in the international market. From the perspective of Taiwan, the policy issue was where Taiwanese investment funds should be directed. On the one hand, protectionism was on the rise in the US and Europe. On the other hand, it would cause concerns for national security if Taiwan's investment overseas became too dependent on

China. Thus, Southeast Asia was chosen as the appropriate target for Taiwan's investment funds to flow overseas. However, since the end of the 1980s the wave of Taiwan investment overseas has gradually moved toward China. This is mainly because of the success of China's economic reforms, the softening of the hostility across the Straits, and the terms that ASEAN countries offered to Taiwanese investors, which became less attractive than before.

The implication of Taiwan's investment swing to China lies in the fact that Taiwan, for the first time, has the opportunity to trade and cooperate with the three superpowers simultaneously. Based on its common interest with Japan and the US, membership in various organizations in the Asia-Pacific region, and its massive capacity in national defence, Taiwan not only stands at the core of the triangle formed by the three superpowers, but also has the opportunity to integrate its economy with China via trade and investments. This affords Taiwan the possibility to affect the path of China's economic reforms after Deng. On the other hand, it also leads Taiwan to the central position of an economic circle formed by Chinese populations. The implication of the latter is that regional economic integration has become one of the most spectacular phenomena of the world economy in the 1990s. Therefore, on the basis of the above, it is reasonable to assume that Taiwan's bargaining power to claim itself as a separate political entity under the one-China concept is greater than before.

Taiwan's current economic policy toward China is "easy out but not easy in." In other words, exporting to China and investing in China is much less restrictive than importing from China. With regard to investment in China, Taiwan's policy is to prohibit investors from investing in infrastructure. This is because such investment projects are largely financed through the public budget. Suppose that China's public budget is allocated among different uses with perfect substitution, then allowing Taiwanese investors to participate in infrastructure investment projects may in turn induce the Chinese government to transfer funds to other uses, including those not favourable to Taiwan's interest. For example, the Chinese government may use the funds thus saved on activities designed to isolate Taiwan diplomatically. Secondly, with regard to imports from China, Taiwan adopts a policy that allows imported goods or raw materials that satisfy the following conditions: those that are non-threatening to Taiwan's security or its local industries, and those having a positive effect on the competitiveness of Taiwan's exports. In general, with the exception of national security, Taiwan's trade policy toward China is mercantilist in spirit, and lays down the fundamental constraints for both sides across the Straits to further integration and specialization.

Fundamentally, from the viewpoint of the US, Taiwan's current trade policy toward China is in accordance with the American strategic objective against China since the end of the Cold War. The premise of the argument is that the formation of a common strategic and economic alliance with Taiwan to countervail China is in the US interest. This argument originates from the change in the US attitude toward China in the post-Cold War era. According to Klintworth (1995), although the US and China are not enemies, they are not as intimate as they were. Also, there have been some conflicts between them over such issues as human rights, the extension of China's status in trade with the US as the most favoured nation, and intellectual property rights. As a result, the US not only supports Taiwan's pursuit of its membership in WTO and APEC, but also promotes its contact with Taiwan to higher official levels.

However, as the chapter has pointed out from the outset, if China uses the hostage issue against the US as it did during the missile crisis, the US government would be forced to negotiate more seriously with China over the Taiwan issue. That is to say, as compared to before, the set of political activities in which Taiwan would be allowed to participate internationally and officially would be smaller. Under such circumstances, one wonders which framework would be most appropriate for Taiwan to consider in devising its optimal reaction in so far as trade policy toward China is concerned.

The Model

Assume that a country's ultimate objective is to maximize the national interest of its own people, say π. Suppose national interest has three main components: economic interest, political interest, and the costs required to maintain any given strategy profile of the three players, that is, α, β, and C respectively. Assume further that the relationship between economic interest and political interest is multiplicative in form as shown by (1) below:[3]

$$\pi_i = \alpha_i \cdot \beta_i - C_i \quad \text{for } i = \text{A}, \text{B}, \text{T} \qquad (1)$$

where A, B, and T stand for Washington, Beijing, and Taipei respectively.

For the sake of simplicity, I assume that Japan's national interest is in line with that of the US. Therefore, we can focus on the triangular relationship among the three players without Tokyo.

Define X_A as the degree of support from the US to Taiwan in the arena of international politics and diplomacy, and X_B as the degree to which Beijing endures Taiwan's participation in that arena.[4] Define Y as the degree of openness of Taiwan's trade policy toward China. Since the three variables, X_A, X_B, and Y represent the choice and policy indices of Washington, Beijing, and Taipei respectively, it can be assumed that the ranges of these variables all lie between zero and one, as shown by (2) below.

$$0 < X_A, X_B, Y < 1 \qquad (2)$$

Equation (3) defines China's national interest function.

$$\pi_B = \alpha_B (X_A, X_B, Y)\, \beta_B (X_A, X_B) - C_B (X_A, X_B) \qquad (3)$$

The signs of each derivative of α_B, β_B, and C_B with respect to X_A, X_B, and Y are assumed by (4).

$$\frac{\partial \alpha_B}{\partial X_A} > 0, \frac{\partial \alpha_B}{\partial X_B} > 0, \frac{\partial \alpha_B}{\partial Y} > 0, \frac{\partial \beta_B}{\partial X_A} < 0, \frac{\partial \beta_B}{\partial X_B} < 0, \frac{\partial C_B}{\partial X_A} > 0, \frac{\partial C_B}{\partial X_B} < 0, \quad \text{.......... (4)}$$

The meaning of most derivatives in (4) is very straightforward. However, $\frac{\partial C_B}{\partial X_B} < 0$, needs more explanation. It suggests that if Beijing allows Taipei more room to pursue its identity as a formal member in the international community, the cost to Beijing will be lower given the degree of the support of the US to Taiwan, that is, X_A.

For Beijing, the policy choice variable is X_B. If Washington is in an opposite position to Beijing over the Taiwan issue, then Beijing's optimal decision would be to maximize π_B given any combination of X_A and Y. Its first order condition is shown by (5).

$$\beta_B \frac{\partial \alpha_B}{\partial X_B} + \alpha_B \frac{\partial \beta_B}{\partial X_B} = \frac{\partial C_B}{\partial X_B} \quad \text{.......... (5)}$$

Implicit in (5) is that there is a reaction function of X_B to any profile of (X_A, Y), which is denoted as X_B^R as shown by (6).

$$X_B^R = X_B^R (X_A, Y) \quad \text{.......... (6)}$$

Similarly, from the standpoint of the US, its national interest function can be expressed by (7).

$$\pi_A = \alpha_A (X_A, X_B) \beta_A (X_A, X_B, Y) - C_A (X_A, X_B) \quad \text{.......... (7)}$$

The signs of each derivative of α_A, β_A, and C_A with respect to X_A, X_B, and Y are assumed by (8).

$$\frac{\partial \alpha_A}{\partial X_A} < 0, \frac{\partial \alpha_A}{\partial X_B} > 0, \frac{\partial \beta_A}{\partial X_A} > 0, \frac{\partial \beta_A}{\partial X_B} > 0, \frac{\partial \beta_A}{\partial Y} < 0, \frac{\partial C_A}{\partial X_A} > 0, \frac{\partial C_A}{\partial X_B} < 0, \quad \text{.......... (8)}$$

Comparing (8) with (4), there are two significant differences.

The first is $\frac{\partial C_A}{\partial X_A} < 0$. This suggests that it would be disadvantageous to the economic interest of the US to raise its support level to Taiwan. The implication of this assumption is that Taiwan will be given a stronger incentive to negotiate with the US on economic issues once it is given more flexibility to pursue its own identity. Secondly, the role which Y plays is asymmetric in π_B and π_A. Y appears to be an endogenous variable of α_B, but it is not an endogenous variable of α_A. Instead, it is an endogenous variable of β_A. The reason for such a formulation lies in the idea that, as put forward by Klintworth (1995), Taiwan's openness to trade with China will benefit China economically, and will in turn help China to be stronger. The latter is, however, not favourable to the long-term political interest of the US.

As Washington and Beijing are in a hostile state, the optimal policy of the US is to choose X_A such that π_A can be maximized given the values of X_B and Y. The first order condition and the reaction function of X_A implied by it, denoted as $X_A{}^R$, are expressed in (9) and (10). The meaning of both equations is similar to that of (5) and (6), and will not be addressed here.

$$\beta_A \frac{\partial \alpha_A}{\partial X_A} + \alpha_A \frac{\partial \beta_A}{\partial X_A} = \frac{\partial C_A}{\partial X_A} \quad \text{.......} (9)$$

$$X_A{}^R = X_A{}^R (X_A, Y) \quad \text{.......} (10)$$

Finally, for Taiwan, its national interest function can be shown as (11).

$$\pi_T = \alpha_T (X_A, X_B, Y)\ \beta_T (X_A, X_B, Y) - C_T (X_A, X_B, Y) \quad \text{.......} (11)$$

The signs of each derivative of α_T, β_T, and C_T with respect to X_A, X_B, and Y are assumed by (12).

$$\frac{\partial \alpha_T}{\partial X_A}, \frac{\partial \alpha_T}{\partial X_B}, \frac{\partial \alpha_T}{\partial Y}, \frac{\partial \beta_T}{\partial X_A}, \frac{\partial \beta_T}{\partial X_B} > 0; \quad \frac{\partial \beta_T}{\partial Y} < 0, \quad \frac{\partial C_T}{\partial X_A} < 0, \quad \frac{\partial C_T}{\partial X_B} < 0, \quad \frac{\partial C_T}{\partial Y} > 0, \quad \text{.......} (12)$$

In equation (12), the sign of ∂βT/∂Y is assumed to be negative. This is because I assume that the more deliberate the trade policy toward China, the more serious the hostage issue facing Taiwan would be, especially where Taiwanese investors in China are concerned. Given X_A and X_B, Taiwan's best strategic reaction is to choose Y to maximize its own national interest. The first order condition and the resultant reaction function are shown by equations (13) and (14).

$$\beta_T \frac{\partial \alpha_T}{\partial Y} + \alpha_T \frac{\partial \beta_T}{\partial Y} = \frac{\partial C_T}{\partial Y} \quad \text{.......} (13)$$

$$Y^R = Y^R (X_A, X_B) \quad \text{.......} (14)$$

If we solve for the reaction functions of the three players simultaneously, that is, equations (6), (10), and (14), we will reach a Cournot-Nash equilibrium, which is denoted by (15).

$$(X_A{}^C, X_B{}^C, Y^C) \quad \text{.......} (15)$$

The above solution can only give us a very obscure impression. It says that the solution will depend on the functional form of each country's national interest under the assumption that each player adopts Cournot behaviour. Certainly, Cournot behaviour is not consistent with what we have observed from the actual pattern of interactions among the US, China, and Taiwan. To model their behaviour more realistically, some simplifications are required. For simplicity, suppose that the national interest functions of the three players are outlined as (16) to (18). I wish to emphasize equation (18). In setting up the national interest function of Taiwan, I assume that Taiwan is pro-US, that is, its national interest, to a large extent, is dependent on the support given by the US.

$$\pi_A = \alpha_A(X_A)\,\beta_A(X_A, Y) - C_A(X_A) \quad (16)$$

$$\pi_B = \alpha_B(X_B, Y)\,\beta_B(X_B) - C_B(X_B) \quad (17)$$

$$\pi_T = \alpha_T(X_A, Y)\,\beta_T(X_A, Y) - C_T(X_A, Y) \quad (18)$$

The signs of the derivatives of a_i and b_i (for i =A , B , T) with respect to X_A, X_B, and Y, where available, are shown in (19) to (21) with each corresponding to (16) to (18) respectively. Basically, these signs are the same as those appearing in (4), (8), and (12) with similar reasoning.

$$\frac{\partial \alpha_A}{\partial X_A} < 0,\ \frac{\partial \beta_A}{\partial X_A} > 0,\ \frac{\partial \beta_A}{\partial Y} < 0,\ \frac{\partial C_A}{\partial X_A} > 0 \quad (19)$$

$$\frac{\partial \alpha_B}{\partial X_B}\ ,\ \frac{\partial \alpha_B}{\partial Y} > 0,\ \frac{\partial \beta_B}{\partial X_B}\ ,\ \frac{\partial C_B}{\partial X_B} < 0 \quad (20)$$

$$\frac{\partial \alpha_T}{\partial X_A}\ ,\ \frac{\partial \alpha_T}{\partial Y}\ ,\ \frac{\partial \beta_T}{\partial X_A} > 0,\ \frac{\partial \beta_T}{\partial Y}\ ,\ \frac{\partial C_T}{\partial X_A} < 0;\ \frac{\partial C_T}{\partial Y} > 0 \quad (21)$$

Next we consider two situations. The first situation simulates the triangular interactions among the three players before the 1996 missile crisis, while the second simulates the interactions after the crisis, especially after Clinton's visit to China. In the first situation, the fundamental assumption is that the US and Taiwan are in strategic alliance, but their relationship is asymmetric in the sense that the US acts like the leader while Taiwan like the follower. In this case, the US will be in a position to internalize Taiwan's reaction function to maximize its own national interest. Nevertheless, the behaviour of the US will not violate the common interest of its alliance with Taiwan since, politically, Taiwan's interest is contingent upon the support of the US.

In the second situation, however, negotiations between the two superpowers are brought formally on to the table. The structure of the game becomes a repetitive one with several stages, each involving two rounds. In the first round, Washington and Beijing make a deal to maximize their joint interest. After that, Taiwan makes its best adjustment at the second round. As Taiwan makes its move, the game moves to the next stage since the US and China will have to negotiate again after Taiwan's move. To be sure, the adjustment mechanism described above is somewhat arbitrary, since the two big players may have the power to internalize Taiwan's reaction function, and force the game to end earlier. Having said that, it would be implied that Taipei does not have its own independent trade policy toward China. I consider the credibility of this argument to lie in the extent to which the two big players can trust each other. If they can work together, it will leave Taiwan very little room to make its own policy. However, the independent existence of Taiwan even without a formal identity internationally suggests that Taiwan has its own role to play, which is unlikely to be a "dependent" one.

Next, I will focus on the analysis of the two situations, in particular the determination of equilibrium in each situation, its properties, and the differences between the two equilibria.

Situation I: Game Before the Missile Crisis

In this situation, the game is characterized by the fact that the US and Taiwan form an alliance, and the US acts as the leader.

As Taiwan is pursuing its maximum interest, we can deduce from (18) that the best reaction of Y, denoted as Y^R, will be the function of X_A.

$$Y^R = Y^R(X_A) \quad (22)$$

In the first stage, Washington substitutes Taipei's reaction function, that is, (22), into the interest function of the US. The optimal value of X_A, denoted as X_A^*, can be solved from the first order condition. Substituting X_A^* into (22), we then have Taipei's optimal choice of Y, denoted as Y*. Therefore, the optimal policy coordination of Washington and Taipei can be described by (23).

$$(X_A, Y) = (X_A^*, Y^*) \quad (23)$$

With regard to Beijing, because of its antagonistic position with the US over the Taiwan issue, its best policy is to choose X_B, while taking Y* as given, to maximize equation (17), as shown by (24).

$$X_B = X_B^* \quad (24)$$

Together with (23) and (24), we then have the equilibrium strategic profile of the triangular relationship, that is (X_A^*, X_B^*, Y^*) before the missile crisis. An important point here is that there is no prior information that X_A^* and X_B^* will be equal. The reason for such an inequality to prevail is the great discrepancy in opinion between the US and China over the Taiwan issue, which results in a lack of mutual communication with each side making choices to maximize self-interest. On the other hand, given the assumption that both the US and Taiwan share a common political interest, as demonstrated by $\frac{\partial B_A}{\partial Y} < 0$, $\frac{\partial B_T}{\partial Y} < 0$ we will have the following inequality:

$$X_A^* > X_B^* \quad (25)$$

It needs to be emphasized that the equilibrium strategic profile is contingent on the functional forms of the three players' national interests, above all on the relationships of parameters, whether complements or substitutes, in the political interest, economic interest, and cost functions of each country. I will not pursue the issue further. Readers can think of the actual relationships as the top secrets in the minds of each player.

SITUATION II: AFTER CLINTON'S VISIT TO CHINA IN 1998

In this situation Beijing cooperates with the US to maximize their joint interest. After that, they have to negotiate on the terms to divide the net gains. To eliminate the difference between X_A^* and X_B^*, both parties will have to take a roughly equivalent or common position, either in favour of, or against, Taiwan in order to benefit from their cooperation.

Suppose that both the US and China assume that Taipei will move according to (22) predicted in the first round of each stage of the game. To reach an agreement on X_A or X_B, denoting the agreed value as X*, such that the joint interest of the two big powers can be maximized, then X* must fulfill the following condition, that is (26):

$$X^* \in \text{Argmax}$$
$$\alpha_A(X)\,\beta_A(X,Y^*) + \alpha_B(X,Y^*)\,\beta_B(X) - C_A(X) - C_B(X) \quad \text{.......} \quad (26)$$

Since XA* and XB* as shown in (23) and (24) are solved while Beijing and China are in a hostile relationship, in situation II it would violate the interest of the US if X* is smaller than XB*. By the same token, it would not be in Beijing's interest if X* is greater than XA*. As a consequence, the inequality below must hold true if the system is solvable.

$$X_A^* > X^* > X_B^* \quad \text{.......} \quad (27)$$

The idea of (27) is very clear: the US will have to make concessions in its support of Taiwan to reduce its costs while increasing its own economic interest at the price of its political interest. For Beijing, the opposite scenario applies, however. Therefore, the agreement between both parties over the Taiwan issue can be seen as an exchange of economic interest for political interest.

It can be noticed from (26) that the outcome of the negotiation at the first stage will not be the equilibrium. This is because as X_A^* has adjusted to X* but $X^* < X^*_A$, then according to (22), Y* will no longer be Taiwan's optimal response. Thus, Taiwan will have to adjust its choice of Y as suggested by (22). This adjustment will serve Taiwan's interest for two reasons. First, it ensures the maximization of Taiwan's interest. Secondly, it will have an impact, though indirectly, of encouraging the US and China to renegotiate over the Taiwan issue. One would expect that the structure of the game is one of repetition with each stage involving two rounds. The equilibrium will be reached once no players have the incentive to change their choices given the choices of the others. During the dynamic process for finding the equilibrium, Taiwan will be in the saddle path, which is composed of all the points of Y^R as X changes along with the two big powers' renegotiation. That is to say that Taiwan will locate at the balance path formed by the two forces.

The implications for Taiwan of these findings are that it is crucial to maintain flexibility in its economic and trade policy toward China, and to be able to manage its capital movements across the Straits. On the other hand, Taiwan's economic and trade policy would become less flexible if both Taiwan and China acquired memberships in the WTO. A potential field for economic policy interactions will emerge once the markets across the Straits have become more easily accessible to both sides.

Concluding Remarks

This chapter discusses Taiwan's economic and trade policy relating to China in the framework of game theory in which three main players are involved. The fundamental assumption upon which the model is built is that the mode of interaction among the big powers surrounding Taiwan, including the US, Japan, and China, have evolved into a coordination game due to the historical evolution in the Asia-Pacific region in the past. To simplify the model, this chapter treats the US and Japan as a single group, and focuses on the triangular interactions among the US, China, and Taiwan.

The change in the mode of interaction among the players occurred in 1996 when China launched missiles over the sea near Taiwan. The supposition is that China's use of the hostage issue against the US was intended to force Washington to negotiate with China directly over the Taiwan issue. Before 1996, the structure of the game was characterized by the hostile relationship between the two big powers, while the US and Taiwan were united in alliance. That alliance was asymmetric, being dominated by the US. After the missile crisis, above all after Clinton's visit to China, the structure of the game changed to a repetitive one characterized by several stages with each stage involving two rounds. At each stage, the US and China negotiate over the Taiwan issue in the first round, while Taiwan makes its own move accordingly in the second round.

For the present, Taiwan needs to maintain its flexibility in dealing with China to realize its potential maximum interest. As is predicted from the model, the best bet for Taiwan is to use the negotiation between the US and China over the Taiwan issue to serve its goal of moving in the direction of independent existence. However, this does not imply that Taiwan has to tighten its economic links with China. The key lies in the fact that Taiwan has to take into account the impact of these links upon the interest of the US. It is suggested that Taiwan may have to adjust its current policy of refraining from being impatient when dealing with China to a more flexible policy.

Endnotes

1 For example, it can be considered as in the political interest of the US to support an independent and democratic Taiwan. By contrast, the military costs that might thus occur, or the additional costs which might be incurred by China's sabotage of US trade were the US involved in cross straits warfare, can be seen as economic concerns.

2 I assume that neither side enjoys absolute advantage over the case of Taiwan. To maximize their joint interest, this negotiation involves an exchange between political gains and economic gains.

3 This implies that the substitution rate between economic interest and political interest is diminishing. Conceptually, α is similar to the sum of principal and returns on safe assets, while β is similar to that on risky assets.

4 In the extreme case where X_A equals one, we can think of it as the situation where the US supports Taiwan's independence.

References

Fewsmith, J. (1994). *Dilemmas of reform in China: Political conflict and economic debate.* New York: M.E. Sharpe, Inc.

Klintworth, G. (1995). *New Taiwan, New China—Taiwan's changing role in the Asia-Pacific region,* New York: St. Martin's Press, Inc.

Lin, J. and Lo, C. (1998). *Zero-sum or win-win? A reinterpretation of Cross-Strait economic exchanges.* Taipei: Academic Sinica.

Schive, C. (1995). *Taiwan's economic role in East Asia.* The Center for Strategic and International Studies, Washington.

Wendell, L., and Lilley, J.R. (1994). *Beyond MFN: Trade with China and American interests.* Washington, DC: The AEI Press.

Plate 13 Defunct cracker company ➧

CRACKERS
MITZ
可口莉滋
...there
is only one
Mitz!

11

China's Future Economic Development: Regionalization

C.W. Kenneth Keng

Associate Professor and Director, Asia Pacific Research and Development, Faculty of Business, University of Victoria

Introduction

One of the most striking economic developments in the late 20th century has been the two-decade long rapid growth of the Chinese economy. A sustainable growing economy is a necessity for China's future stability. Sustainability of that growth depends on China's continued commitments to institutional reform and economic deregulation. Relaxation of government intervention in economic activities has led, and will lead, China to decentralize central government authority over economic planning and control. This will stimulate the emergence of regional economies in mainland China. In the next two decades, there will likely be 10 regional economies with relatively independent industrial structures emerging in Greater China[1] as a result of economic decentralization.

If the central authority follows the laws of the free market and leaves regional economies intact, these regional economies are likely to experience robust productivity, which will prolong the current rapid growth into the first two decades of the new century. China's neighbouring economies, Taiwan and Hong Kong (together with Macao), may also contribute to and benefit from this anticipated trend of economic regionalization if they integrate their industrial structures and economic markets with the mainland's economic regions, so that the maximum benefit from scale economies will be enjoyed.

This chapter first presents three potential prospects for China's economic growth up to 2020: high-growth potential based on the policy of economic deregulation and decentralization, stable growth potential based on an extrapolation of current policies and trends, and a low growth scenario postulated on a reversion to an authoritarian government. Under the stable growth scenario, China will progress along lines of gradualism or incrementalism by "groping for stones on which to cross the river." This will enable the economy to grow at a compound annual rate of 5.8% between 1997 and 2020. Under the socialist hard-line option, the economy will grow much more slowly at an annual rate of 4.3% in the same period. If China speeds up its efforts toward economic deregulation and decentralization, it will achieve its high growth potential at an annual rate of 7.4% up to 2020. Should this take place, by 2020 the average Chinese mainlander will enjoy a living standard as high as Taiwan residents enjoy at present.

This chapter then proceeds with an investigation of the limitations to China's future economic growth and a discussion of the emerging economic regionalization on the Chinese Mainland. Benefits of integrating the Taiwan and Hong Kong economies with neighbouring Chinese regions are also discussed and a projection of the integrated Chinese economies in the new century is presented. The chapter closes with an outlook for the Chinese regional economies in the new century.

China's Economic Growth Prospects

There are various predictions for China's future economic growth. In its official development plans, the Ninth Five-Year Plan and the Fifteen-Year Perspective Plan, the PRC's State Council outlined an explicit short-term target for an average annual GDP growth of 8% through the period of 1996 to 2000 and a long-term growth prospect of 7% per year up to 2010. Most of the economic growth predictions published in China do not deviate to any noticeable degree from the official "forecasts."

With the aid of a multi-sector neoclassical growth model, the World Bank projected China's long-term average annual growth rates of GDP at 8.4% for 1996-2000, 6.9% for 2001-2010, and 5.5% for 2011-2020 (World Bank, 1997a). These predictions can be computed to obtain a compound growth of 6.6% per annum for the period of 1996 to 2020, which is about 0.6% per year lower than China's official plan. The World Bank's prediction, as clearly stated in its report, was simply an extrapolation of China's past growth experience (mainly from the 1980s to the early 1990s) into the future. It did not incorporate such qualitative information as the depth of institutional reforms due to the political sensitivity of the subject. As well, the World Bank has been prohibited from publishing any detailed economic forecasts about China or any other member countries. Nevertheless, the World Bank's study did provide many model-estimated or model-simulated economic parameters that are invaluable in appraising China's growth prospects. These growth prospects for the Chinese economy can be summarized as: a high annual savings rate of 40% combined with high total factor productivity advancement of 2% per year that may produce an annual economic growth of 7.9%; or a relatively low annual savings rate of 20% together with low productivity growth of 1% per year that would result in an annual growth of 4.2% for the Chinese economy as a whole (refer to Table A-1 of the Appendix). These simulated prospects are based on the World Bank's multi-sector neoclassical (Solow) growth model. Its parameters are summarized in Table A-2 of the Appendix.

Based on the prescribed World Bank long-run steady-state multi-sector growth model of China, Keng (1998) incorporates many qualitative assumptions concerning China's demographic changes, pace and depth of institutional reforms, degree of decentralization and progress in terms of regionalization, productivity advancement, savings rates, as well as the international business environment (e.g., membership in the World Trade Organization) to generate three principal scenarios as China's potential economic growth prospects up to 2020. These scenarios are based upon the presumptions that for the entire forecasting horizon up to 2020 there will be no large scale conflicts or severe impediments in world trade and investment systems, international affairs will be conducted as usual under the rulings of the UN and the WTO, and China will have a relatively stable social and political environment. These three possible economic prospects for China's economic growth are: 1) the stable growth scenario based on an extrapolation of China's current policies and trends; 2) the low growth scenario postulated on a reversion to an authoritarian socialist government in China; and 3) the high growth potential based on the policy regime of economic deregulation and decentralization.

The Stable Growth Scenario

If Beijing maintains the current pace in its pragmatic approach to China's future economic development (that is, the central authority gradually decentralizes its power over economic affairs and incrementally relaxes its interventions and controls over markets at a steady pace similar to what was observed in the reform era of the 1980s and early 1990s), the Chinese economy will grow steadily at an average (compound) annual rate of 5.8% through the next two decades (refer to the Stable Growth path in Figure 11.1). Under this *Business as Usual* scenario, China's economy will produce US$3,137 billion (at 1996 price levels) of goods and services annually by 2020; or equivalently, the size of the Chinese economy will grow to 3.9 times of that in 1996. Since China's population will still grow at a positive rate during this period (refer to Keng, 1996b), China's GDP per capita will be expected to grow at a compound rate of 4.9% per year. This will allow an average Chinese citizen to earn, in 2020, an annual income of US$2,153 at 1996 price levels. If the ratio of China's prices to international prices remains unchanged, China will have a GDP per capita of US$10,136 at the 1996 purchasing power parity (PPP). This average income of US$10,136 is about the same as that of Malaysia, or about 86% of that of South Korea's in 1996.[2]

The Low Growth Scenario

If China reverts to its past way of governing, that is the authoritarian approach of strong central planning with intense government intervention, alters its reform and open-door pro-growth development policies and reinforces strict socialist philosophy and policies, the Chinese economy will grow at a much slower pace of 4.3% per year between now and 2020. Under this *Socialist Hardline* scenario, intensive central planning and extensive government intervention will retard the market mechanism, and result in overbearingly inefficient allocation of resources and high economic transaction costs. People's propensities to work and save will largely deteriorate, and both domestic and international investments will be devastated. The Chinese economy will probably experience economic growth that is primarily generated merely by its population growth and some slow accumulation of capital. Consequently, by 2020, China's economic scale will be merely 2.7 times that in 1996 (refer to the Low Growth path in Figure 11.1), or merely two-thirds of that under the stable growth scenario. An average Chinese individual will earn about US$1,509 per year (at the price levels) in 2020, or equivalently, US$7,104 at the1996 PPP. This income level is approximately equivalent to what Mexicans had in 1996.

The High Growth Scenario

The high growth scenario postulates that China, in the next decade, will speed up its institutional reforms and relax most of its central government intervention in regional economic affairs; and, in order to acquire adequate materials and energy resources to sustain its rapidly growing economy, China will have to accelerate its globalization process to accommodate a more liberal division of production, better capital movement, and freer international trade. Under this *decentralization and deregulation* prospect, regional economies with relatively independent industrial structures will emerge vigorously as a result of scale economies and market competition. Free regional markets and competitive regional industries will attract more international investment and technology transfers. Meanwhile, Chinese regions will actively invest abroad to secure their international sources of energy and materials as well as to sustain international markets for their products. This will lead to more efficient use of capital and labour at lower transaction costs than under the other scenarios. Hence, higher growth is expected.

Under the high growth scenario, it is anticipated that the Chinese economy will grow at an average rate of over 7.4% per annum up to 2020 (refer to the High Growth path in Figure 11.1). This will bring the Chinese economy to a size 5.6 times as large as in 1996. The total value of goods and services produced in the economy in 2020 will reach US$4,552 billion computed at the 1996 price level. Equivalently, China's GDP per capita in 2020 will reach US$3,133 at 1996 prices, or US$14,752 at the 1996 PPP. This personal income level is even higher than Taiwan's at $14,295 in 1996. This means that if China further deepens its institutional reform and continues to open its markets, market forces, scale-economy impetus, and industry-structural optimization will keep accelerating its process toward regionalization and productivity advancement. By 2020, China will have advanced to one of the world's largest industrialized economies with the average living standard that Taiwan enjoys today.

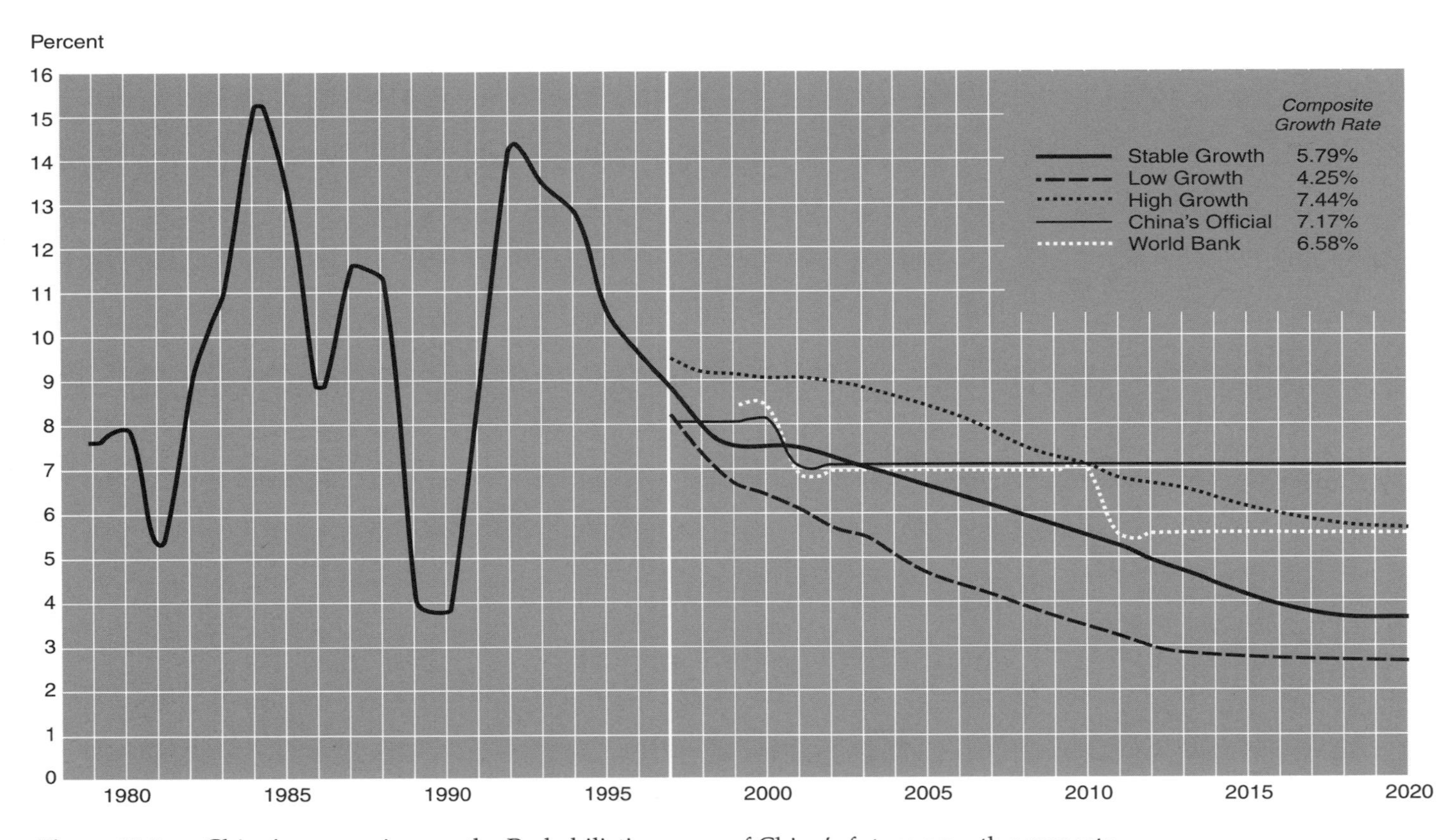

Figure 11.1 China's economic growth: Probabilistic ranges of China's future growth prospects

Figure 11.1 demonstrates these predictive scenarios. Interested readers are referred to Keng (1998) for a detailed discussion. It can be seen from Figure 11.1 that throughout the forecasting horizon up to 2020, the World Bank's prediction is, on average, about 0.78% higher than our prediction under the Stable Growth scenario, and China's official plan is, on average, about 1.38% higher than the Stable Growth prediction. However, the other predictions are below our predictions under the High Growth Scenario.

Accounting for all possible outcomes based on these simulations using the parameters generated by the World Bank's model, Keng (1998) suggests that the prior probability[3] for China's future economic growth being between the High Growth and the Low Growth scenarios is about 80%. There is a 10% chance that China's future growth will either exceed the predicted High Growth path or fall below the Low Growth path. In other words, the long-run trend of China's economic growth in the next quarter century will most likely, with an 80% degree of confidence, fall between the High Growth and the Low Growth predictions. Meanwhile, the Stable Growth path may serve as a representative scenario (or in forecasting jargon, the most-likely scenario) for China's future economic growth.

The Limitations to China's Growth

China possesses more than one-fifth of the world's population but only 7% of the world's arable land. In terms of its resource base per capita, China is among the most poorly endowed countries in the world. The World Bank's 1992 study showed that China's per capita agricultural land is 28% of the world average; range lands per capita are less than half the world average; forests and wilderness areas per capita are only 15% of the world average; water resources are about one-third the world average; and energy resources per capita, with the exception of coal, are also very low. China's poorly endowed natural resources could severely limit its economic growth and welfare improvements in the future, unless more efficient means (e.g., better technology) of utilizing these resources are adopted. Although less government intervention, internationalization of markets, and globalization of industrial structures would enable China to resolve a large portion of these natural resource constraints,[4] the relatively low per capita endowment of plains remains a critical and tenacious physical restriction on China's economic growth.

When demand for urban land rises with increasing urbanization and industrialization, land prices or cost of using land services tend to rise relative to other prices. The high cost of plains together with government regulations on urban land use will impede the effort of the Chinese economy to build a national transportation system. The limited supply of plains land for urban and rural housing, commerce, and industry, will thus be a crucial physical limitation to China's future economic development. The high imputed cost of using plains land is a critical challenge to China's capital productivity and international competitiveness. Therefore, how to utilize urban land intensively rather than extensively is the key bottleneck for China's future economic growth.

The Shortage of Plains

China is endowed with an area of 9.6 million km^2, 12% (or 1.15 million km^2) of which are plains. According to a recent satellite survey, China has about 1,988 million mu[5] (1.32 million km^2) of cultivated land,[6] which accounts for about 13.8% of its total area. Of the cultivated land, 1,290 million mu (859,269 km^2) are on plains (with a gradient not exceeding 8 degrees) accounting for 75% of China's plain area. This leaves only 25% of China's plains (290,000 km^2) for urban, industry, land transportation development, infrastructure construction, and other non-agricultural uses.

China's per capita area of plains for non-agricultural use in 1996 was 238 m^2. In the next two decades, when China's population grows to 1.5 billion, this per capita statistic will "shrink" to 194 m^2 if China maintains its agricultural use of plains to the extent it does today (Table 11.1). On the other hand, if in the next quarter century China keeps its present standard of per capita non-agricultural use of plains, its agricultural use of plains will have to be reduced in order to accommodate the growing urban population and the huge increases in industrial and commercial developments. In this case, by 2020, China will have to have released an area as much as 40% of the current area of agricultural plains for non-agricultural uses! Needless to say, this will bring all kinds of problems, such as inadequacy of domestic food supply, impaired environment, and so on. Chinese strategists have considered these problems as priority national security issues. It seems that in the near future there will be little chance that Chinese authorities will lower the announced food sufficiency level below 90% (the current level is 95%). Therefore, there is a severe limitation on the release of China's agricultural land for other uses.

Table 11.1 China's use of plains: 1996 vs. 2020

Year	*Population (million)*	*Plains (1,000 km^2)*	*Per capita plains (m^2)*	*Per capita plains for non-agricultural use (m^2)*	*Plains for non-agricultural use (km^2)*
1996	1,223.89	1,152	941	238	290,811
2020	1,502.00	1,152	766	194*	356,893+

Original Data Source: *China Statistical Yearbook 1997.*

* Based on the 1996 plains area for non-agricultural uses.

+ Based on the 1996 per capita plain area for non-agricultural uses.

In 1992, Japan had a per capita plains area of 792 m^2, and in the US each person on average enjoyed over 25,000 m^2 of plains. China's per capita plains area in 1997 was 941 m^2 and is expected to decline to 767 m^2 by 2020 when its population reaches 1.5 billion. If the span (extensiveness) of land is taken into account, it should immediately be seen that China's land transportation use of plains should be much larger than Japan's. This is because China's area is 25.3 times Japan's and only one third of Chinese provinces are coastal. Therefore, China requires a more extensive long-distance land transportation system than Japan does. Land transportation, due to construction technology, capital and operation costs, as well as the location of population centres, should mostly take place on the plains. Heavy demand for long-distance land transportation means that more highways and railroads have to be built on China's relatively rare plains.

Transportation Use of Plains

In 1997, China had 133,339 km of railways and 1,185,789 km of highways. If, on average, each kilometre of railroad occupies 0.004 km^2 of land and each kilometre of highway occupies 0.005 km^2 of land,[7] the land required for transportation totals an area of 56,299 km^2. China's per capita transportation land in 1997 can thus be estimated at 46.1 m^2. This is half as large as Japan's 90 m^2 in 1992. Furthermore, if 80% of China's land transportation is on the plains, it occupied 45,039 km^2 of China's plains in 1997, which is over 3.8% of China's total area of plains.

Between 1996 and 2020, China will probably have to accommodate a population increase of 278 million (see Table 11.1). If it is conservatively assumed that China will not increase its per capita transportation use of plains in the next quarter century, an additional area in plains of over 10,000 km^2 is still needed for building new land transportation capacity to accommodate the extra economic activities generated by those additional 278 million people. By 2020 the total area of plains used for land transportation would be over 55,000 km^2, which is 4.8% of the national total (see Table 11.2). This estimate may serve as the lower limit of what China requires for land transportation by 2020, because it is based on China's actual experience at present. Twenty years from now, the Chinese economy will expand to a size at least three times as big as it is today. The much larger economy will generate much more demand for domestic land transportation. As a result, per capita transportation use of land tends to increase significantly.

Table 11.2 also lists alternative estimates based on different assumptions of per capita use of land for land transportation. It will cost 9.4% of its plains if China will have constructed a land transportation system by 2020 similar to that of Japan in the 1990s. As an island nation, Japan has an area merely one twenty-fifth that of China, and unlike China, Japan imports most of its required energy, natural resources, and materials to the major industrial regions located along its coast. This means Japan's demand for long-distance land transportation is relatively low. In contrast, 90% of China's required energy and natural resources and 95% of its required grains are at present domestically produced. Even in the remote future, when China fully participates in the international division of production, it will still have to produce the vast majority of goods and services it needs domestically.

Given China's geographically unbalanced distribution of natural and energy resources, the future demand for land transportation should be significantly larger than that of Japan at present. We tend to agree with Wang (1997a) that it is appropriate to use 1.5 times Japan's per capita transportation use of land (that is, 135 m^2 per capita) as the basis to estimate that of China in 2020. This brings the estimate to 155,250 km^2 of total area of land needed for China's future land transportation, which is about 1.6% of the national area and 14.1% of the national plains (refer to the third row of figures in Table 11.2). There is another alternative estimate based on the present experience of the US. It would use up more than half of its plains if China would build a system of highways and railroads similar to the one Americans enjoy today (refer to the bottom row in Table 11.2).

Table 11.2 Estimates of China's land transportation use of plains

Year or Country	*Land used for land transportation (per capita)*	*Land used for land transportation (national total)*	*Plains used for land transportation (national total)*	*Transportation share of total plains*	*Additional area of plains needed for transportation*
1996	46 m^2	56,299 km^2	45,039 km^2	3.9%	-
2020⁺	46 m^2	69,000 km^2	55,200 km^2	4.8%	10,161 km^2
2020*	135 m^2	202,500 km^2	162,500 km^2	14.1%	117,461 km^2
Japan	90 m^2	135,000 km^2	108,000 km^2	9.4%	62,961 km^2
US	496 m^2	744,000 km^2	535,200 km^2	51.7%	490,161 km^2

Original Data Source: *China Statistical Yearbook 1997* & Wang (1997a).

⁺ Based on China's actual status in 1996.

* Based on 1.5 times Japan's status in 1992.

US & Japan figures are from Wang (1997b).

Urbanization Use of Plains

If China's degree of urbanization reaches 60%by 2020, there will be 900 million people living in urban areas. Using Tokyo's 50 m^2 per capita urban use of land in 1994 as the benchmark, China's urban use of land will be 45,000 km^2 (Table 11.3). This is equivalent to about 3.5% of China's plains area.[8] Suppose that 70% of the rural population's houses are built on the plains—rural housing then requires 34,500 km^2 of plains in 2020. China's total urbanization and rural housing needs for plains can then be estimated as 61,000 km^2, accounting for 6.5% of China's total area of plains (refer to the second row in Table 11.3). This estimate is essentially the most conservative because it is based on statistics from the world's most densely populated metropolitan centre, Tokyo, with extremely high land prices.

An alternative estimate is based on Japan's national average per capita urban use of land. If all the cities are taken into account, Japan's per capita urban use of land was 115 m^2 in 1992. Using this as the standard, by 2020, China will need 172,500 km^2 of land for urban development, that is 1.8% of China's total area of land and 15.0% of China's total plains. If 90% of urban cities are on the plains, the actual use of plains for urbanization by 2020 will be 155,250 km^2, accounting for about 1.6% of China's total area and 13.5% of plains (refer to the first row in Table 11.3).

Table 11.3 also lists estimates based on the average urban use of land in Shenzhen and Beijing. At present, Shenzhen is the only metropolitan area in China that was built according to the principles of modern urban planning. However, Shenzhen has so far accommodated mainly light industries and is still far from being fully commercialized. Therefore, its per capita urban use of land tends to fall short of that of a fully industrialized and commercialized metropolitan area. Consequently, Shenzhen is not a representative city to be used as the benchmark for a typical Chinese metropolitan area in the future. Neither are Beijing and Shanghai, which have so far been mainly symbolized by their "old glories." The vast majority of their streets, houses, and business areas were planned over 50 years ago. There are very few areas of these two cities that have been planned and rebuilt under modern urban planning. Hence the statistics for Beijing and Shanghai seriously underestimate the future land requirement for an average Chinese metropolitan centre.

Table 11.3 Estimates of China's urban and rural housing uses of plains in 2020

Alternative models	*Urban use of land in m^2* (per capita)*	*Urban use of land in km^2 (national total)*	*Urban use of plains in km^2 (90% of the total urban use of land)*	*Urban share of plains (%)*	*Rural housing use of plains in km^2*	*Rural share of plains (%)*	*Total share of plains (%)*
Japan average	115*	172,500	155,250	13.5	34,500	3.0	16.5
Tokyo	50*	45,000	40,500	3.5	34,500	3.0	6.5
Shenzhen	30*	27,000	24,300	2.1	34,500	3.0	5.1
Beijing	20*	18,000	16,200	1.4	34,500	3.0	4.4

* Figures in this column are cited from Wang (1997a).

China's central planning agency, the State Planning Commission, estimates that by 2010 the degree of urbanization will reach 60%, and urbanization will occupy 1% and transportation will occupy 1.3% of its total area—assuming China adopts Japan's "metropolitan model" to develop its economy. That is, China would not, and could not, plan for a nation-wide division of production, because a nation-wide industrial structure similar to that of the US requires extensive use of land transportation in order to move resources, materials, intermediate and final goods, and personnel. China physically does not have enough plains land to build even a minimal capacity of land transportation to accommodate a nation-wide industrial structure for a population of 1.5 billion.

Even if China adopts one-half of the US standard on a per capita basis to build its land transportation system, such a system would still take up more than 25% of its plains. That means there will be severe conflicts in demand for plains space for environmental protection, food production, transportation, and urbanization. It will be even more devastating if China's poor per capita reserve of petroleum resources are taken into account because petroleum fuels will remain the principal fuel for both highway and railroad transportation in the foreseeable future.

The physical limitation attributed to its low per capita endowment of plains will impel China to diversify its national economy into regional ones. Each of China's regional economies should essentially have its own relatively independent industrial structure and market system, so that cross-region long-distance land transportation is economized, and henceforth the use of plains for transit systems is largely reduced. By doing this, transportation costs which constitute an important portion of the transaction costs of almost all economic activities can be minimized. The Chinese economy as a whole will therefore use less resources for transportation to produce the same level of outputs. This essentially provides a strategic and economic rationale for China to decentralize its resource allocation power to its regions. It is indeed a more efficient strategy for China's future economic growth. Productivity gains from this strategy of economic regionalization have been estimated to be as high as 1.5% per year on average through 2020 (Keng, 1999). In effect, if there had been no central planning, and China had relied completely on market forces to allocate its resources for economic development, economic regionalization would have already been a virtual reality in China.

China's Economic Regionalization

Taking into consideration China's history, geographic-economic characteristics, and current state of regional economic development, many studies tend to suggest that in the first decade of the 21st century there will be a minimum of nine regional economies emerging in Mainland China (Wang, 1997a, 1997b;

Keng, 1998, 1999). The emergence of these regional economies may be partially attributed to China's current economic decentralization and market deregulation. When the central authority relaxes its control over regional economic decision making, market forces will guide economic activities toward establishing lower-transaction-cost industrial structures and market systems so that regional economies will emerge. Under a market mechanism, resources can be intensively and therefore efficiently used by being allocated toward urban centres where infrastructure and business services provide economic production units with comparative advantages. As a consequence, individual regional markets and industrial systems, each with a network of closely affiliated cities, will emerge and become relatively independent regional economies within the vast Chinese national geographical area. This process of regionalization is essentially what Japan has experienced in the last 30 years (Wang, 1997a).

Accounting for all possible methods of land utilization at the national level, China's State Planning Commission study (SPC, 1996 & 1997) on regional and industrial development concludes that China will have no better option than to adopt Japan's model of regionalizing its industrial structures if costs for building large scale long-range transportation corridors are to be minimized. The SPC study also calls for faster urbanization to establish metropolitan economies so plains and resources will be used intensively. These metropoli connected with surrounding satellite cities would constitute an urban net on which a synchronized multi-sector industrial system can be strategically planned and developed. Ideally, the metropolitan area is the regional centre of manufacturing industries and commercial services, and satellite cities may house specific industries. The regional industrial structure is strategically planned, so that a major proportion of goods and services produced in the region are consumed within the region itself. This will largely reduce the need for long-range interregional transportation.

In order to maintain sustainable economic growth, China should advance its industrial structure and economic efficiency as rapidly as possible, so that its relatively limited natural resources (on a per capita basis) may be utilized to the maximum extent. To meet the immense demand of the future economy, China will have no alternative but to build up its manufacturing capacity and raise its productivity to compete in the international market in order to acquire needed grains, raw materials, oil and petroleum products, capital equipment, advanced technology, and so forth. However, due to transportation costs (the costs of transit infrastructure construction and operations), limited petroleum resources, inadequate plains for building transit systems and for urbanization, China will be unable to rely on a single national division of production, but instead will have to depend on a multitude of regional systems of production and consumption. This will most likely impel Chinese authorities to adopt vigorous regional economic development policies as a national strategy for economic growth in the new century.

Historical Trends

It has been a "tradition" of the Chinese agrarian economy that large portions of China's goods and services were consumed within regions where they were produced. A World Bank study (World Bank, 1997c) indicates that even in the early 1990s more than two thirds of China's agricultural products were consumed within the local counties and towns where they had been produced. This agrees with our own finding, that interregional economic activities represent a small portion of China's total economic activities even in the late 1990s.

As recently as 1996, 47 years after the Communists consolidated Mainland China, the "national economy" of Mainland China was still a collective of small regional economies with a surprisingly small fraction of interregional economic interactions. As seen in Table 11.4, interregional economic activities measured by long-distance freight and passenger traffic via railways, waterways, and civil aviation were only 22.86% (freight) and 9.85% (passengers) of their national totals. The highway freight and passenger traffic, representing 75.9% and 90.15% respectively of total traffic, have average travel distances of 51 km and 44 km. These statistics indicate that long-distance interregional economic interactions are minimal, even at the very end of the 20th century. Though China has been the largest political entity in the world for the last half century, its national economy still consists of relatively small and autonomous regional economies.

Table 11.4 China's regional economic interactions: Freight and passenger traffic and volume in 1996

	Total	*Railway*	*Highway*	*Waterway*	*Civil Aviation*	*Pipelines*
Freight (10,000 ton)	1,296,200	168,803	983,860	127,430	115	15,992
Share	100.00%	13.02%	75.90%	9.83%	0.01%	1.23%
Average distance (km)	281.24	768.38	50.93	1,401.75	2,167.83	365.81
Volume* (100 million ton-km)	36,454.00	12,970.46	5,011.20	17,862.50	24.93	585.00
Share	100.00%	35.58%	13.75%	49.00%	0.07%	1.60%
Passenger (10,000 persons)	1,244,722	94,162	1,122,110	22,895	5,555	-
Share	100.00%	7.56%	90.15%	1.84%	0.45%	-
Average distance (km)	73	353	44	70	1346	-
Volume* (100 million person-km)	9,143.00	3,325.37	4,908.79	160.57	747.84	-
Share	100.00%	36.37%	53.68%	1.77%	8.18%	-

Data Source: *China Statistical Yearbook 1997.*

* Volume = Traffic x Distance

Another study (Wang, 1997b) shows that in 1994 the volume of inter-regional economic activity between major neighbouring regions was surprisingly small. Using railway shipments as a measure, Wang found that merely 10% of Liaoning's shipments were with the neighbouring Capital Region (consisting of Beijing, Tienjin, and the whole province of Hebei) and the Shandong Region (the peninsular province), and less than 12% of Shandong's inter-regional shipments were with the Capital and Liaoning regions. One of Wang's conclusions was that these three neighbouring regions around the bay area of the Bohai Sea (or the so-called Bohai Bay Area) are large economies (Shandong's and the Capital Region's population are over 85 million and Liaoning's exceeds 40 million) with a comparatively sound stock of natural resources. Therefore, they have developed their own relatively independent and autonomous industrial structures, and enjoyed their own internal markets. Needs for inter-regional shipments of industrial goods, are hence, fairly small. However, we believe that the somewhat minimal inter-regional economic interactions among these three regions around the Bohai Bay may also be the result of a lack of inter-regional transportation links. There have been few inter-regional shipments of industrial goods among these three neighbouring regions. Or, maybe these three Bohai Bay regions have been developed relatively independently with limited dependence on inter-regional economic interactions.

Institutional and Policy Oriented Trends

In the 1990s the Chinese central government has tended to delegate more power and responsibilities to regions. The central state's transference of one of its key powers, grain policy, to provinces in 1995 symbolizes the trend of regionalization in economic decisions. It has always been a key policy objective of China's central government to maintain a nationwide balance of demand and supply for grain, and not many would deny that one of the few economic achievements of the Chinese Communist regime has been "providing food security to one fifth of the world population." However, in 1995 the central government decentralized its food-security responsibility to provincial governments. Under the new "Governors Responsibility System,"[9] provincial governments assume the responsibility to balance their own province-wide grain supply and demand through inter-provincial grain trade. This essentially established an inter-provincial wholesale commodity market of grain to replace centrally planned inter-provincial grain transfers. It also replaced the nationwide grain security policy with a multitude of provincial policies.

Under the decentralized grain security system, provinces are free to trade grains and to subsidize farmers' inputs and consumers' purchases at the discretion of the provinces to the extent that budgets permit. However, provinces must meet the provincial stock responsibility; that is, surplus-producing provinces must maintain a 3-month supply of grain stocks, and deficit provinces a 6-month supply. The state retains the power of collecting the grain quota,[10] the responsibility of maintaining the required national stock, and the international grain trade monopoly. Private traders (including provincial grain enterprises) are not allowed to procure grain in the countryside until state grain quotas have been filled. The consequences of this decentralization of grain security policy is the stimulation of grain production, the liberation of the grain market, and the deregulation of grain prices. Starting in 1997, China's state government has procured grain quotas at "government unified" market price and discontinued consumption subsidies through the grain enterprises.

It has also been anticipated that, due to the large diversity in economic development among regions, China will most likely depend on provinces to institutionalize their own social welfare and pension systems. If these two key pillars of the "socialist market economy," namely the national grain production and trade system and the social security system, can be delegated to provinces, we do not see any reason why other economic decisions cannot be decentralized to regions. For example, China has recently accelerated reform of its banking system by establishing nine regional (multi-provincial) central banks (similar to the Federal Reserve Banks in the US) and many regional commercial banks serving areas across administrative regions (such as provinces). Furthermore, provinces have already shouldered the responsibility for all primary and secondary, and a large portion of post-secondary education. China's latest plan for education reform, as announced in its National Education Conference in June 1999, calls for the "sale" of a large portion of traditionally state-financed universities and colleges to provinces. This means the production of human capital in the future will be mainly under regional responsibility.

Once Chinese economic regions have their own discretion to exercise economic sovereignty over food supply, social welfare, human resources, and so forth under the framework of a national market (or even in an international market), different regional social-economic systems are expected to emerge on the Chinese mainland. That is, economic regionalization may eventually lead to the emergence of plural social (even political) sub-systems in China.

Nine Emerging Regional Economies

If China's central authority keeps its current trend of market-based institutional reform and steadily decentralizes its economic power to regional authorities, many regional economies with relatively autonomous industrial structures and markets will emerge as a result. If Chinese macroeconomic strategic planners take into account the natural resource constraints on China's future growth, and accelerate the process of economic decentralization and regionalization, this process of regionalization may even be accelerated. Wang (1996) and many other studies (e.g., Keng, 1997) have suggested that it is likely that there will be 9 to 10 regional economies with metropolitan centres emerging in China in the coming one to two decades. This possibility essentially falls under the "decentralization and regionalization" scenario of China's future growth pattern introduced in the last section. Should this happen, China will probably enjoy its highest growth potential, namely a compound annual growth over 7% up to 2020.

Table 11.5 lists the nine metropolitan regions and their statistics. These nine economic regions together comprised 67% of China's population, 74% of the plains, and 71% of the GDP in 1996. Among them, five are coastal regions, including: Liaoning, the Capital and Shandong regions located around the Bohai Sea on China's north coast; Greater Shanghai, comprising the vast area of the Yangtze River delta and the Qientang River delta and occupying about a quarter of China's coastline; and the Pearl River Delta Region located at the southern coast. The remaining regions include one near-coastal region, West Jiangsu and Anhui; and three inland regions, Jilin-Heilongjiang at the north-east corner neighbouring with North Korea and Russia; Hunan-Hubei-Jiangxi (or the South-Central Region) at the centre of the Mainland; and Sichuan in China's midwest.

At present, all of these regions have built, or are building, fairly modern intra-regional highway systems and they have all either developed their own regional markets with relatively independent industrial structures, or are in the process of doing so. Inter-regional transportation relies heavily on frequently congested railways. These inter-regional transit systems are far from adequate to meet the shipping needs of a nationwide industrial system, and due to constraints of the physical geography, it is not only very expensive to build modern inter-regional transit systems, but at many locations it is almost impossible.

These emerging regional economies are a natural consequence of China's urbanization and marketization. Their development pattern is surprisingly similar to that of Japan's. Japan's regional economies typically have a radius of 200 to 300 km (or an area of 126 thousand km^2 to 283 thousand km^2) with populations over 30 million. Each of these regional economies is characterized by a network of urban centres where a relatively independent structure of industries is located. The urban network is a group of geo-economic affiliated cities containing one or two metropolitan areas as manufacturing and commercial services centres complemented by a number of smaller satellite cities that may specialize in specific industries. Connected together by well-developed inter-urban transit systems, the metropolitan centres and satellite cities with their surrounding rural areas constitute a relatively autonomous production and consumption system, or, a regional economy.[11] Market forces impel the allocation of regions' resources toward constructing localized industrial structures, so that maximum production and consumption, and therefore the best economic benefits, are enjoyed. Inter-regional activities are engaged in only when comparative advantages exist.

Although the central government may use macroeconomic policies through a national fiscal (taxation) system to transfer resources inter-regionally and to ensure national interests, most inter-regional economic exchanges should rely

Table 11.5 Nine emerging metropolitan economies in China—1996

Major Regional Economies	*Population (million) 1996 year end*	*Share in total population (%)*	*Area (1,000 km²)*	*Share in total area (%)*	*Per capita plains (m²)*	*Plains (1,000 km²)*	*Share in plains (%)*	*Major metropolitan areas or cities*	*Regional GDP (US$billion)²*	*Share in national GDP (%)*	*Per capita GDP (US$)²*
Capital Region	86.91	7.10	216	2.25	1,070	93.0	8.07	Beijing, Tienjin, Shijiangzhuan	74.261	9.13	854
Liaoning	41.16	3.36	146	1.52	1,214	50.0	4.34	Shenyang, Dalien	37.999	4.67	923
Jilin-Heilongjiang	63.38	5.18	656	6.83	3,679	233.2	20.24	Changchun, Harbin	45.003	5.54	710
Shandong	87.38	7.14	153	1.60	1,122	98.0	8.51	Jinan, Chingdao	71.726	8.82	821
Hunan-Hubei-Jiangxi	163.58	13.37	564	5.88	868	142.0	12.33	Wuhan, Changsha, Nanchang	85.856	10.56	525
Sichuan	114.30	9.34	567	5.91	123	14.0	1.22	Chengdu, Chongqing	50.722	6.24	444
Pearl River Delta	69.61	5.69	186	1.94	603	42.0	3.65	Guangzhou, Shenzhen, Zhuhai	78.449	9.65	1,127
West Jiangsu & Anhui	78.70*	6.43*	191	1.99	915	72.0	6.25	Nanjin, Yangzhou, Hefei	54.227*	6.67*	689
Greater Shanghai	83.78*	6.79*	184	1.92	1132	109.9	9.54	Shanghai, Soozhou, Hangzhou, Ningbo, Wuxi, Changzhou	114.800*	14.12*	1,370
Sub-total	816.52	66.72	2,869	29.8	932	854.1	74.14		579.343	71.26	710
Other regions	407.13	33.26	6,737	70.2	726	297.9	25.86		233.650	28.74	574
China	1,223.89	99.98[1]	9,600	100.0	941	1,152.0	11.98[3]		812.993	100.00	664

Data Source: *China Statistical Yearbook, 1996 &1997. A Strategic Study of China's Regional Development*. State Planning Commission, Beijing, 1996

[1] Military personnel are not included.
[2] 1996 mid-year US$-RMB$ exchange rate = 1: 8.313. Plains share in total area
* Estimates based on provincial data

on the market mechanism. This regional development model of the metropolitan economy is indeed the trend that most Chinese regions have followed, and will most likely follow in the future. It is probably the least costly approach to adopt for China's future economic development given its severe natural resource constraints. To an extent, this trend of regionalization has stemmed from, and will be largely affected by, the central government's policies of gradually decentralizing the economic decisions to regions.

FUTURE CHINESE REGIONAL ECONOMIES: THE CHINESE TIGERS

If China speeds up its process of regionalization, it will enjoy the benefits of higher degrees of market openness and international division of production. Decentralization of economic sovereignty, and regionalization of industrial structures, will create a free environment for inter-regional exchanges and competition. Market competition will guide resources to their optimal allocations and minimize transaction costs. Inter-regional competition will ensure the best possible economic growth for the Chinese economy as a whole. As a consequence, China will not only attract more overseas commercial investment, but will also induce Hong Kong and Taiwan to accelerate their economic integration with Chinese regions at large, and with the neighbouring regions of the Pearl River Delta (Guangdong) and Fujian in particular. These four sub-regional economies will be in a much better position to jointly optimize their industrial structures, capital and labour utilization, and market scale, so that they will attain their maximum economic strength. On the other hand, the rapidly growing regions of Guangdong, Fujian, Hong Kong, and Taiwan, with a joint population of over 101 million in 1996, will generate a huge production capability to support modernization in other Chinese regions. They will also create a tremendous demand for the goods and services of other Chinese regions. China will then enter an era of full-strength growth, as a result of the integration of Taiwan and Hong Kong with the Chinese southeast coastal regions.

Keng (1997 and 1998) has suggested that the southeast coastal province of Fujian (positioned between China's two main coastal metropolitan regions, Greater Shanghai and the Pearl River Delta, with a population of 30.2 million), which has been "left out" in the SPC study of nine metropolitan economies, can be easily (in terms of economic transaction costs) integrated with Taiwan, to become an export-oriented regional economy. Fujian is relatively isolated geographically by mountains and hills from its neighbouring provinces of Zhejiang, Jiangxi, and Guangdong. Also, the Pearl River Delta region, or Guangdong, can be easily integrated with its highly commercialized and internationalized neighbours Hong Kong and Macao, to become another export-oriented economy.

Benefits of Regional Integration

If they are integrated with Fujian and Guangdong, Taiwan and Hong Kong (and Macao) will benefit from higher returns to their existing capital stocks, as well as new investment opportunities, improved scale economies, abundant supplies of primary (less-skilled) labour, food and materials at much lower costs, and expanded markets for their manufactured goods, modern services, and so on. With the same amount of factor inputs, Taiwan and Hong Kong will produce more output if they get freer access to Fujian's and Guangdong's vast markets of resources, commodities, and services. In other words, the real marginal returns to labour and capital in Taiwan and Hong Kong will be rising. As a consequence, per capita GDP (or average real income) will grow faster. These economic advantages are expected to generate, on average, 2% to 4% economic growth in addition to normal growth otherwise anticipated (Keng, 1998).

There are many reasons why Taiwan and Hong Kong will attain much higher productivity if they are integrated with Fujian and Guangdong. If integrated with Fujian, Taiwan will probably restructure its industries so that low productivity sectors such as the agriculture sector will be moved to Fujian where land prices, water resources, and labour costs are much lower. On the one hand, Taiwan's investment in Fujian's agricultural sector will generate a much higher return; and on the other hand, Taiwan's down-sized agricultural sector will release valuable land as well as huge sources of capital and labour to its industrial and commercial sectors.

Labour mobility between Taiwan and Fujian will not only increase the productivity of both economies, but it will also decrease their real unit labour costs. As a result, both economies will enjoy enhanced competitiveness in international markets. If integrated with Fujian, Taiwan's labour market will be restructured. Fujian's abundant and inexpensive labour may adequately meet Taiwan's shortage of less-skilled labour (e.g., construction workers, assembly-line operators, steel mill workers, housemaids, and so forth). Taiwan's well-educated and trained manpower can then be better utilized for work of higher productivity. Meanwhile, Taiwan's professionals and technicians with traditional (labour-intensive) technology and skills may contribute more if they were to engage themselves in Fujian's economic development. This could lower Fujian's required investment in human capital and therefore shorten Fujian's process of training skilled labour and professionals.

Furthermore, not only will Taiwan enjoy lower prices of factor inputs, but in addition it will gain access to the Mainland's markets in food products and raw materials, with much better terms of trade than otherwise. Therefore, the costs of producing and living in Taiwan will be lowered. This will consequently improve the competitiveness of Taiwan's products and services in international markets. At the same time, since Fujian is a part of the Mainland, products of enterprises in Fujian financed by Taiwanese will become "domestic" products with access to all parts of the vast market of China. This will not only increase the market share of Taiwanese enterprises in China, but also enable them to optimize their production scales. Hence, better production efficiency will be enjoyed. The expansion in production and improvement in productivity will generate additional growth for Taiwan's economy. As shown in Table 11.6, these economic advantages are expected to generate 2% to 5% economic growth in addition to normal growth under the status quo.

Table 11.6 Taiwan's additional growth contributed by integration

	Productivity gain (%)	*Additional growth (%)*
Sectoral shift of labour	0.5 - 1.5	0.5 - 1.5
Larger market and scale economy	0.5 - 1.0	0.5 - 1.0
Deletion of tariffs and other trade barriers	0.5 - 1.0	0.5 - 1.0
Lower costs of factor inputs and materials	0.5 - 1.0	0.5 - 1.0
Production division	1.0 - 2.0	0.5 - 1.0
Total		2.5 - 5.5

Source: Keng (1998)

Future Chinese Economic Regions: Chinese Tigers

If integrated with Taiwan and Hong Kong, Fujian and Guangdong will benefit hugely from Taiwan's and Hong Kong's capital, technology, management know-how and modern commercial services. Based on market incentives, their capital investment needs will probably be adequately supplied by Taiwan and Hong Kong. As a result, Fujian and Guangdong will experience another two decades of high growth without consuming China's scarce capital. Once Fujian and Guangdong start to integrate with Taiwan and Hong Kong to become open economies, their industrial structures will evolve toward an industrial and service orientation. This means that there will be another run of structural changes in Fujian and Guangdong: less productive labour will flow to higher paid industries and further economy-wide productivity gains can be enjoyed. Hong Kong's and Taiwan's quality commercial services and international connections will also lead Fujian and Guangdong to the international market at relatively low transaction costs. Consequently, Fujian and Guangdong will not only become more productive, but also more integrated with international markets. Taiwan-Fujian and Hong Kong-Guangdong will become new regional economic powers with economic scales (in terms of GDP) greater than their neighbouring Chinese regions, Greater Shanghai and the South-Central region, which possess significantly larger populations.

Since the required capital and technology for development of Guangdong and Fujian will be sufficiently supplied by Taiwan and Hong Kong, China may direct its scarce capital resources to economic development of other regions. As a result, there will likely have emerged at least four new world-class regional economic powers within Greater China: Taiwan-Fujian, Hong Kong-Guangdong, Greater Shanghai, and the South-Central region. As indicated in Keng (1997 & 1998), in the first two decades of the new century these four regions will experience higher growth than other regional economies in China. Keng (1998 & 1999) also predicts that, by 2020, the Taiwan-Fujian and Hong Kong-Guangdong economies will have passed South Korea in size, and become the new Asian Tigers in the 21st century (see Table 11.7 and Figure 11.2).

Taiwan and Fujian had a combined population of over 50 million in 1996, and a GDP more than one third that of China (Table 11.7). Taiwan has industrialized itself in the 1990s. Its manufacturing capacity and technology, human and financial capital, degree of internationalization, and commercial expertise will sustain its integration with Fujian, where raw materials, less-skilled labour, and natural resources are all relatively more available than in Taiwan.

Hong Kong and Guangdong had a combined population of over 75 million and a total GDP of US$224.6 billion in 1996—more than a quarter of China's GDP. Hong Kong's population of 6.3 million essentially speak the same dialect as the people of Guangdong, and for the last two decades almost all of Hong Kong's manufacturing plants have been moved to the Pearl River Delta area. Hong Kong has become the port city servicing southern China with its advanced financial systems, international commercial expertise, and modern transportation and communication facilities and technology. It is estimated that if fully economically integrated with Fujian and Guangdong, Taiwan and Hong Kong may grow at a rate of 8% and 7% respectively, from the present to 2020. Otherwise, Taiwan and Hong Kong will maintain their status quo for the next two decades, at best growing at 5% and 4% per annum respectively.

Endnotes

1. Greater China means the neighbouring economic area of Hong Kong, Macao, Taiwan, and Mainland China.

2. Based on *Asiaweek* issued on September 26, 1996, the 1996 GDP per capita calculated at PPP of the following countries were in US currency: US - $26,825; Japan - $22,220; South Korea - $11,750; Taiwan - $14,295; Hong Kong - $23,892; Macao - $16,840; Singapore - $23,565; Malaysia - $9,470; Russia - 5,260; Thailand - $7,535; Mexico - $7,188; India - $1,385; and China - $2,935. Since *China Statistical Yearbook 1997* shows that China had a GDP per capita of RMB$5,634 in 1996 and the official RMB$ and US$ exchange rate in July 1996 was 8.31, this equates China's GDP per capita in 1996 to US$678 which is much higher than the GNP per capita of US$540 used in *Asiaweek* (this may be due to differences in base year and in exchange rates adopted). Accordingly, we adjust China's 1996 GDP per capita at PPP to US$3,685. In this study, we use 5.435 as the 1996 PPP factor to adjust China's aggregate price level to the international one.

Table 11.7 "Chinese Tigers" vs. "Asian Tigers": Present and future

Economy	*Population (million)*	*1996 GDP US$ billion (1996 dollar)*	*1996 GDP (PPP) US$ billion (1996 dollar)*	*1996 GDP per capita US$ (1996 dollar)*	*1996 GDP (PPP) per capita US$ (1996 dollar)*	*Compound annual growth rate*	*2020 GDP US$ billion (1996 dollar)*	*2020 GDP per capita US$ (1996 dollar)*
Status Quo Scenario								
Korea	44.8	451.4	526.4	11,750	13,702	5%	1,456	33,793
Taiwan	21.6	264.9	308.8	12,265	14,298	5%	854	35,274
Hong Kong	6.3	146.2	150.5	23,200	23,882	4%	375	58,111
Singapore	3.1	81.8	73.1	26,400	23,592	4%	210	66,127
China	1,223.9	825.4	4,486.2	678.0	3,685	6%	3,187	2,187
Guangdong	68.8	78.4	426.4	1,145	6,222	7%	398	3,692
Fujian	32.61	31.4	170.5	979	5,321	7%	159	3,158
Macao	0.4	7	6.7	17,475	16,726	4%	18	43,771
Economic Integration Scenario								
Taiwan-Fujian	54.2	296.3	479.3	5,407	8,747	8.3%	1,989	30,252
Taiwan	21.6	264.9	308.8	12,265	14,298	8.0%	1,680	69,578
Fujian	32.6	31.4	170.5	979	5,321	10.0%	309	7,745
Hong Kong-Guangdong	75.5	224.6	576.9	2,911	7,475	8.3%	1,514	15,967
Hong Kong	6.3	146.2	150.5	23,200	23,882	7.0%	742	93,935
Guangdong	69.61	78.4	426.4	1,145	6,222	10.0%	773	9,056
Greater Shanghai	83.8	114.8	623.9	1,370	7,446	8.5%	813	7,771
China South-Central	163.6	85.9	466.6	525	2,853	8.5%	608	2,977

Data Source: China Statistical Yearbook 1997, A Study in Regional Development Strategies of China (1996), Asia Week, September 1996.
US$ and RMB$ exchange rate = 1: 8.31 (Mid-year rate)
PPP Price Factor: In 1996, US$1.00 worth of goods and services in China is assumed to be equivalent to US$5.435 (PPP).
1996-2020 Population Growth Assumptions: Taiwan: 0.5%, HK & Macao: 0.1%, China: 1.0%

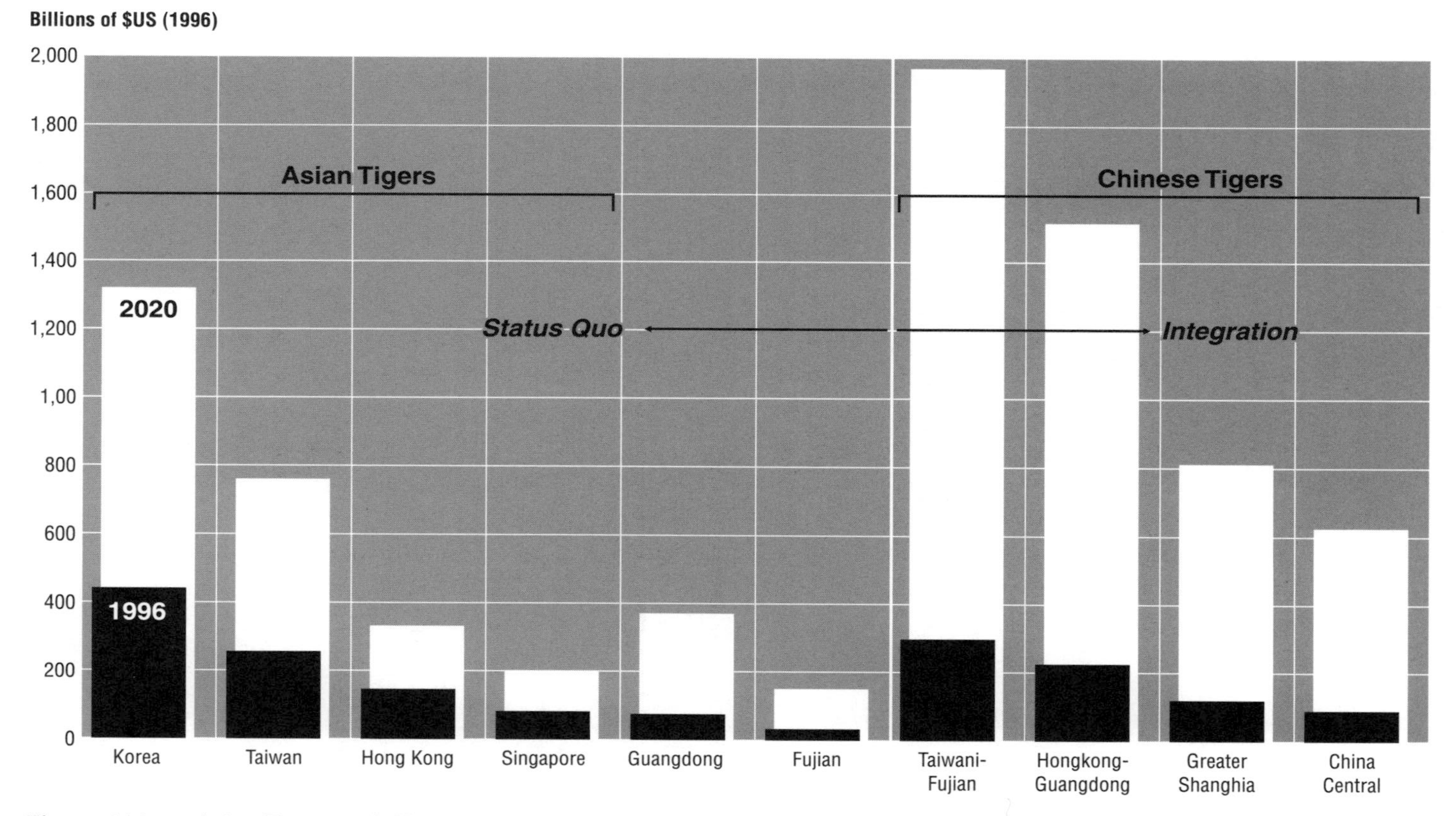

Figure 11.2 Asian Tigers and Chinese Tigers: 1996 vs. 2020

[3] Here we are essentially using the concept of Bayesian subjective probability. In a world of non-repeatable events, any probabilistic assessment attached to a possible outcome of an uncertain event is a statement of the assessor's "degree of belief" and is meaningful only prior to the point in time when that event actually takes place. Once a non-repeatable event takes place, the outcome of that even becomes certain so that there is no more need for any probability assessment.

[4] The World Bank's most recent studies (World Bank 1997c & 1997d) convincingly document evidence that based on world anticipated demand and supply situations in the next quarter century, China's shortage of agricultural and energy resources will not significantly affect international energy and food prices in the foreseeable future, provided that China fully engages in globalizing its consumption and production.

[5] mu is a traditional Chinese measuring unit for land area. 1 mu = 666.1 square meters = 7,200 square feet = 0.164 acres.

[6] China's cultivated land area based on its official statistical survey is 949,700 km^2 in 1996. This is significantly smaller than that based on China's satellite survey of 1.32 million km^2 by as much as 28%. In *China Statistical Yearbook 1997*, it is noted that the statistical survey may underestimate the actual area of cultivated land. In this study, we adopt the satellite survey data because statistical survey data tend to underestimate the actual area due to the obvious reason that China's agricultural taxes are essentially based on its official (statistical) survey of cultivated land. Farmers and local governments have all kinds of reasons to report lower cultivated areas of their land.

[7] It is assumed here that railroads occupy, on average, a 40-metre wide strip of land, and highways occupy a 50-metre wide strip of land. This assumption is essentially of "Chinese characteristics." For example, in the US, the width of a strip of land occupied by a modern highway is at least twice as much as assumed here.

[8] Here we are also assuming that 90% of China's urban land is on the plains.

[9] The governor's responsibility system is less popular than its nickname: "the Rice Barrel Project." Under this system, every provincial governor is personally responsible for the balance of demand and supply of grain (or rice) at the wholesale level in his or her own province. And each city or county mayor is personally responsible for the supply and demand of secondary foods (vegetables, meats, oil, eggs, sugar, etc.) in the retail markets within his or her jurisdictional territory. This "Mayor's Responsibility System" also has a popular nickname of the "Vegetable Basket Project."

[10] Provincial quotas of grain must be delivered to state grain bureaus at designated locations at a lower-than-market price.

[11] In North America, regional economic planners usually use the distance of 500 to 600 miles, or equivalently, the 10-hour (one working day) highway shipping range, as a rule of thumb for site selections for new industrial and commercial units. For example, the site selection of an automobile assembly plant must consider the distances from the engine plants and major parts suppliers.

References

Asian Development Bank (1998). *Statistical database*, February. Manila: Asian Development Bank.

Borenztein, E., and Ostry, J. (1996). Accounting for China's growth performance. *American Economic Review Papers and Proceedings*, 86(2), 224-228

Cai, F. (1996). *Long-term urbanization and key urban policy issues in China*. Washington DC: World Bank.

Chun, D. (1996). China's regional development and economic development zones. In J. Wang (Ed.), *China's economic development strategy and international cooperation* (pp. 12-32). Beijing: China Macroeconomic Society, State Planning Commission. March 1996.

Chun, D. (1997). China's industry development policy and fiscal and monetary policies. In J. Wang (Ed.), *The role of foreign capital in China's industrial and regional development* (pp. 42-54). Beijing: China Macroeconomic Society, State Planning Commission.

Coe, D, Helpman, E., and Hoffmaister, A. (1994). North-South R&D Spillovers, *IMF Working Paper WP/94/144*, Washington, DC: International Monetary Fund.

Colchester, N., and Buchan, D. (1990). *Europower: The essential guide to Europe's economic transformation in 1992*. London: Economist Books.

Ghosh, A. R., and Ostry, J. D. (1997). Macroeconomic uncertainty, precautionary saving, and the current account. *Journal of Monetary Economics*, 40(1), 121-139.

Hill, C. W. L. (1997). *International business: Competing in the global marketplace*, 2nd edition, Chapter 8. Chicago: Richard D. Irwin.

Hu, Z., and Khan, M. S. (1996). Why is China growing so fast? *IMF Working Paper 96/75*, the Research Department, Washington, DC: International Monetary Fund.

Keng, C. W. K. (1996). China's land disposition system. *Journal of Contemporary China*, November 1996.

Keng, C. W. K. (1997a). *China's regional economic development: Status, strategy and prospects*, research monograph, Rotman School of Management, University of Toronto.

Keng, C. W. K. (1997b). China's economic prospects: A prediction, Working Paper, Rotman School of Management, University of Toronto.

Keng, C. W. K. (1998). An economic China: A win-win strategy for both sides of the Taiwan Strait. *American Journal of Chinese Studies*, V(2), 182-215.

Keng, C. W. K. (1999). China's economic prospects in the new century. In A. J. Nathan, Z. Hong, and S. R. Smith (Eds.), *Dilemmas of reform in Jiang Zemin's China* (pp. 171-212). Boulder: Lynne Rienner Publishers.

Lardy, N. (1992). *Foreign trade and economic reform in China, 1978-1990*. Cambridge: Cambridge University Press.

Lin, J.Y. (1992). Rural reforms and agricultural growth in China. *American Economic Review*, 82(1), 34-51.

Sachs, J., and Warner, A. (1995). Economic reform and the process of global integration. *Brookings Papers on Economic Activity 1*, Washington, DC: Brookings Institution.

SPC (1996). *China's economic development strategy and international cooperation*, Jian Wang (Ed.). Beijing: China Macroeconomic Society, State Planning Commission.

SPC (1997). *The role of foreign capital in China's industrial and regional development*, Jian Wang (Ed.). Beijing: China Macroeconomic Society, State Planning Commission.

Swann, D. (1990). *The economics of the common market*, 6th edition. London: Penguin Books.

The Economist (1994a). The European union: A Survey, October 22.

The Economist (1994b). Happy ever NAFTA, December 10 .

UNCTAD (1995). *World investment report 1995*. Geneva: United Nations Conference on Trade and Development.

Wang, J. (1997a). China's macroeconomic strategy of regional economic development. In J. Wang (Ed.), *The role of foreign capital in China's industrial and regional development* (pp. 1-19). Beijing: China Macroeconomic Society, State Planning Commission.

Wang, J. (1997b). Regional development issues in China's Ninth Five-Year Plan. In J. Wang (Ed.), *The role of foreign capital in China's industrial and regional development* (pp. 20-24). Beijing: China Macroeconomic Society, State Planning Commission.

Wang, P., and Wu, Y. (1996). Learning from the EC: The implications of European economic integration for China and Taiwan. *American Journal of Chinese Studies*, III(2), 205-224.

Wang, Z., and Tuan, F.C. (1996). The impact of China's and Taiwan's WTO membership on world trade: A computable general equilibrium analysis. *American Journal of Chinese Studies*, III(2), 177-204.

Wei, S. (1996). Foreign direct investment in China. In T. Ito and A. Krueger (Eds.), *Financial deregulation and integration in East Asia* (pp. 97-119). Chicago: University of Chicago Press.

Woo, W. T. (1995). Chinese economic growth: Sources and prospects, University of California at Davis, Department of Economics (unpublished).

World Bank (1992). *World development report 1992: Development and the environment*. New York: Oxford University Press.

World Bank (1994). *China: GNP Per Capita*. Report 13580-CHA, Washington, DC: World Bank.

World Bank (1997a). *China 2020: Development challenges in the new century*. Washington, DC: World Bank.

World Bank (1997b). *Global economic prospects and the developing countries*. Washington, DC: World Bank.

World Bank (1997c). *At China's table: Food security options*. Washington, DC: World Bank.

World Bank (1997d). *Clear water, blue skies: China's new environment in the new century*. Washington, DC: World Bank.

World Bank (1997e). *Old age security: Pension reform in China*. Washington, DC: World Bank.

WTO (1995). *Regionalism and the world trading system*. Geneva: World Trade Organization.

Appendix

World Bank's Quantitative Model of the Chinese Economy

The World Bank's model of the Chinese Economy is a long-run steady-state multi-sector neo-classical (Solow) growth model. The model contains three sectors: Agriculture, Industry, and Services. Each sector is assumed to have a constant-returns-to-scale production function with a fixed factor input, two variable factor inputs of capital and labour, and a variable input of intermediate goods. Unity elasticity of substitution between factor inputs and intermediates together with perfect competition in all markets is also assumed. These assumptions essentially qualify the adopted production technology as long run. Goods and services so produced are either used as intermediate inputs into the production of other goods and services, or for final consumption and investment.

On the consumption side, household aggregate demands for goods and services are derived from maximizing a constant-elasticity-of-substitution (CES) utility function of all the final goods and services subject to the budget constraint of total output minus total savings. The demand for investment goods is also assumed as a CES technology of individual investment goods subject to the budget constraint of aggregate savings. With further assumptions concerning initial conditions and friction in the labour market, the World Bank team managed to obtain the following parameterization (Tables 11.A1 and 11.A2).

Table 11.A1 China's long-term economic growth prospects (annual percentage GDP growth)

	Total Factor Productivity Growth		
Savings-GDP Ratio	1.0%	1.5%	2.0%
20%	4.2%	4.9%	5.5%
25%	4.8%	5.5%	6.4%
30%	5.4%	6.1%	7.2%
35%	5.9%	6.6%	7.6%
40%	6.4%	7.2%	7.9%

Source: World Bank (1997a)

Table 11.A2 The World Bank long-run growth model of China: Parameters

	Agriculture	*Industry*	*Services*
Output Elasticity			
Fixed Factor	.10	.00	.00
Capital	.20	.20	.27
Labour	.34	.09	.24
Intermediates	.36	.71	.49
Intermediate Elasticity			
Agriculture	.39	.10	.30
Industry	.44	.72	.56
Services	.17	.19	.42
Investment Shares	.04	.88	.08
Consumption Shares	.25	.35	.40

Source: World Bank (1997a)

Plate 14 Women in a crafts class, near Pingtung, Taiwan ➧

12

Female Labour Force Participation in Taiwan and Its Effect on Labour Supply

Peter C. Lin

Professor, Graduate Institute of Economics, National Sun Yat-sen University

INTRODUCTION

Making the most efficient use of human resources is always a major concern in economics. An ample supply of labour is vital to economic growth. The rate of female labour force participation in Taiwan increased from 42.1% in 1951 to 45.6% in 1998, which was relatively lower than that of industrialized countries. Relatively speaking, female labour is more flexible than male labour in terms of the entry to and exit from the labour market. The abundant pool of working-age women is a potential source of labour supply that can be mobilized to accommodate the increasing demand for labour by industries.

Many economists have taken a great interest in the study of the major determinants of female labour force participation. Earlier work was based on cross-section data, applying the least-squares method to ascertain important variables affecting female labour force participation (Chang, 1978). The logit model, designed specially for use with a discrete dependent variable, has also been adopted as the estimation method instead of OLS (Lo, 1986), as has the econometric technique of cointegration (Brooks, 1991). The method of time-series analysis has been employed to detect whether or not there is a lead-lag relationship between variables. Specifically, the Granger causality test is used to find out the nature of causality between female labour force participation and other related variables (Kao and Chen, 1994).

Due to labour shortages and difficulties with the recruitment of workers in Taiwan, the intention of this study is to find ways of bringing more females into the labour force. Thus, it is important to have a proper understanding of the sources of variation in female labour participation. However, due to inadequate sources of data and/or different handling of variables, conclusions from past studies on female labour force participation are not reliable. In addition, it is regrettable that the effect of changes in female labour force participation rate on labour supply in Taiwan has not been examined.

In this research, the Household Income and Expenditure Survey and the Labour Force Survey are used as two separate sources of micro cross-section data. The logit model is adopted as the estimation method. The results from these two sources of data are analysed and evaluated for consistency.

A labour supply function is estimated by OLS. Based on the estimated equation, a simulation is performed to establish a baseline prediction of labour force. By changing the female labour force participation rate, a new prediction of labour force is generated. The difference between the baseline prediction and the new prediction corresponding to a change in female labour force participation rate is taken as a measure of its effect on labour supply.

In the next section, trends in female labour force participation are briefly explained. In Section 3, the logit model is introduced as an estimation method. Empirical results of the logit model are presented in Section 4. In Section 5, a labour supply function is estimated to analyse the effects of changes in female labour force participation rate on labour supply. Section 6 performs factor analysis of changes in labour force participation. Finally, in Section 7, we summarize the study and present some conclusions.

Trends in Female Labour Force Participation

After the Second World War, with the return of soldiers and the baby boom thereafter, the labour force participation of women declined continuously from 42.12% in 1951 to 32.54% in 1966. The decline was undoubtedly related in part to the increase in the burden of childcaring due to persistent high fertility together with a fall in infant mortality. The population pressure on land and lack of employment opportunity also contributed to the decline. The labour force participation rate of females rebounded in 1967 and the rate rose to 39.3% in 1982. The rate has remained persistently in the neighbourhood of 45.0% since 1985. The improvement in the job market due to rapid economic growth in the 1980s and the upgrading in education of females contributed partly to the increase in female labour force participation.

During the same period, the labour force participation rate of males had fallen. In 1951, the rate for males was 90.0% and it declined to 78.0% in 1978. It fell continuously and reached its lowest level of 70.6% in 1998. Trends toward earlier retirement among males account for some of the decline, along with changes in other demographic and socioeconomic factors.

The female labour force participation rate in Taiwan was higher than that of Canada prior to 1965. The pattern has been reversed since 1965, with the Canadian rate surpassing that of Taiwan. The gap between the two rates widened over time with a difference of over 12% since 1988. The relatively lower participation rate for females in Taiwan is attributable to differences in socioeconomic factors and policy measures between the two countries. The following are some of the more important factors. First, the oriental tradition of devotion to family after marriage always encourages women to quit their jobs and leave the job market. Second, raising children and household duties are considered the major responsibilities of women in Taiwan. Married women are in fact tied to the family, making them less likely to enter the labour market. Third, lack of childcare facilities also contributes to lower female labour force participation. Fourth, lack of placement services and employment information is another reason explaining the lower female participation rate in Taiwan.

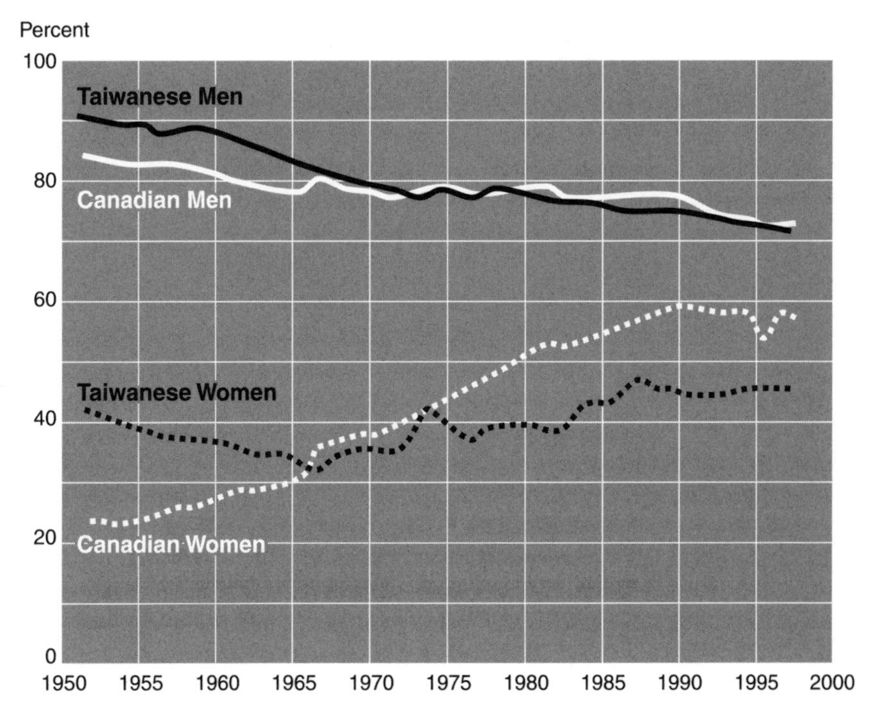

Figure 12.1 Labour force participation rates by sex, Taiwan and Canada, 1951-1997

THE LOGIT MODEL

We will consider the following representation to express the relationship between the female labour force participation rate and its explanatory variables:

$$P_i = E(Y_i = 1 \mid X_i) = \frac{1}{1 + e^{-(\beta_0 + \beta_1 X_{1i} + \cdots + \beta_k X_{ki})}} \quad \text{(1)}$$

where X is a vector of explanatory variables that contains all the characteristics of the worker and Y is the female labour force participation that can take only two values: 1 if the person is in the labour force and 0 if she is not.

Letting $Z_i = \beta_0 + \beta_1 X_{1i} + \cdots + \beta_k X_{ki}$, equation (1) can be written as

$$P_i = \frac{1}{1 + e^{-Z_i}} \quad \text{(2)}$$

Equation (2) represents the logistic distribution function. It is easy to verify that as Z_i ranges from $-\infty$ to $+\infty$, P_i ranges between 0 and 1 and that P_i is nonlinearly related to Z_i (that is, X_i).

If, in equation (2), P_i represents the probability of female labour force participation; then $(1 - P_i)$ is the probability of females not being in the labour force, that is:

$$1 - P_i = \frac{1}{1 + e^{Z_i}} \quad \text{(3)}$$

Therefore, we can write

$$\frac{P_i}{1 - P_i} = \frac{1 + e^{Z_i}}{1 + e^{-Z_i}} = e^{Z_i} \quad \text{(4)}$$

If we take the natural log of equation (4), we obtain the following result:

$$L_i = \ln\left(\frac{P_i}{1 - P_i}\right) = Z_i = \beta_0 + \beta_1 X_{1i} + \cdots + \beta_k X_{ki} \quad \text{(5)}$$

Equation (5) is called the logit model. As P goes from 0 to 1, the logit L goes from $-\infty$ to $+\infty$.

For estimation purposes, we write equation (5) as follows:

$$L_i = \ln\left(\frac{P_i}{1 - P_i}\right) = \beta_0 + \beta_1 X_{1i} + \cdots + \beta_k X_{ki} + \varepsilon_i \quad \text{(6)}$$

It can be shown that if the number of females (N_i) at each X_i level is fairly large and if each observation in a given class X_i is distributed independently as a binomial variable, then

$$\varepsilon_i \sim \text{iid } N\left[0, \frac{1}{N_i P_i (1 - P_i)}\right]$$

that is ε_i follows the normal distribution with zero mean and variance equal to

$$1 / [N_i P_i (1 - P_i)]^1$$

Since the disturbance term in the logit model is heteroscedastic, the OLS is not applicable. We will have to use the weighed least squares in estimating the logit regression.

Empirical Results of the Logit Model

The logit $L = \ln\left(P_i / (1 - P_i)\right)$

is the log of the odds ratio—the ratio of the probability that a female will enter the labour force to the probability that she will not enter. In the estimation of the logit model, we rely on three samples. Sample 1 is taken from the October 1998 Labour Force Survey, Directorate-General of Budget, Accounting and Statistics (DGBAS), Taiwan. Sample 1 contains in total more than 60,000 persons. After screening to exclude male respondents and females aged 65 and over, sample 1 covers 25,949 females. Sample 2 is taken from the 1997 Household Income and Expenditure Survey, DGBAS, Taiwan, by selecting households with females aged 15 to 64. Sample 3 is a sub-sample of the 1997 Household Income and Expenditure Survey. It is chosen to find out the effect of the husband's income on female labour force participation and includes households with both spouses present. This sample contains 6,972 households.

Based on logit estimate results (Tables 12.1-12.3), the following findings are most significant and deserve special attention:

1) Urbanization has a negative effect on the odds ratio of female labour force participation. Relative to rural areas, females residing in cities or towns have a lower rate of labour force participation. The higher cost of day care in cities or towns may discourage females from entering into the labour market. In rural areas, day care is easier to find at a relatively lower cost. In addition, it is much easier for females in the rural areas to find seasonal work or to work on farms during the busy season.

2) Age has a positive effect on labour force participation for females in general; its effect rises at the early stage of their life reaching the highest level at age 36-45 and then declines thereafter. In the early years of marriage, young couples have to make a greater effort to save money for children's education or for a down payment on a house. Thus, women are encouraged to enter the labour market at the early stage of their life to share part of the economic burden of the family. As they reach middle age, the children are grown up and they are better off financially. There is less need for females to join the labour force and, therefore, the labour force participation for females declines gradually after middle age.

3) Education plays the most significant role in labour force participation of females. As years of schooling rise, the odds ratios increase progressively. When years of schooling reach 13 and over, the effect on labour force participation becomes the highest.

4) Past working experience has a positive effect on labour force participation of females. People with work experience undoubtedly have a greater chance of finding a job. Therefore, experienced females are more likely to participate in the labour market.

5) Marital status also exerts a certain effect on labour force participation of females. Relative to married women with spouses deceased, women not married exhibit the highest odds ratio for participation in the labour force, women married with spouses present rank second, and divorced women have the lowest ratio. The unmarried, mainly aged 20 to 30, have a responsibility to be self-supporting and therefore have a high incentive to stay in the labour market. Married women with spouses present may have to stay home to take care of children and, as a result, their willingness to participate in the labour force is somewhat moderated. Divorced women receive either divorce settlements or alimony and the need for them to work may not be as strong as other groups, making their labour force participation relatively low.
6) Household income plays a minor role in affecting the odds ratio. The coefficient is significant at the 1% level, but its magnitude is low.
7) Females residing in the same household as parents-in-law have the highest odds ratio of participating in the labour force. Parents-in-law can help in taking care of children and thus reduce the cost of day care which raises the incentive of females to enter the labour market.

In order to make predictions of participation rates corresponding to different attributes of women, the conditional predictions have been made combining five different sets of variables: urbanization, age, education, work experience, and marital status (Table 12.4). Based on sample 1, the estimated results show that women between 36 and 45 years old with a college education and work experience have the highest probability of participation in the labour market. This probability is higher when the woman is not married, and lives in a rural area. As expected, education has a positive effect on participation and married women participate less than other groups.

Effects of Changes in Female Labour Force Participation Rate on Labour Supply

An increase in a country's working age population (occurring perhaps because of a higher birth rate or immigration) increases the aggregate amount of labour supplied at any given current real wage. The labour force participation rate of this population group affects the level of labour force supplied.

The real wage is an important determinant of labour supply. Higher current real wages raise aggregate labour supply for two reasons. First, when the real wage rises, people who are already working may supply even more hours, for example, by offering to work overtime, by changing from part-time work to full-time work, or by taking on a second job. Second, a higher real wage may entice some people who are not currently working to enter the labour force; that is, a higher real wage may lead to increased labour force participation.

Therefore, we can formulate the labour supply function as follows:

LS = F(RWG, MLFPR, FLFPR, WAP, GRGDP.1) .... (7)

where LS is labour supply, RWG is the real wage, MLFPR is the male labour force participation rate, FLFPR is the female labour force participation rate, WAP is the working-age population, and GRGDP.1 is the previous period GDP growth rate.

By using quarterly data from 1978 to 1998, a labour supply function is estimated as follows:

LS =	-70431.0250	+ 36.355MLFPR	+ 45.794FLFPR
	(-34.962)	(5.047)	(13.256)
	+ 4.076GRGDP.1+	7410.057ln(WAP) +	277.503ln(RWG) .... (8)
	(2.300)	(39.814)	(6.073)
R^2 =	0.999		

Figures in parentheses are t-ratios. All the coefficients of the explanatory variables have the correct signs and are all significantly different from zero at the 5% level, indicating that they all have positive effects on labour supply. The fit of the model is very good as indicated by the high R^2 value.

We would like to use the estimated labour supply function to make predictions of labour supply over the period from the first quarter 1999 to the fourth quarter 2003. For all explanatory variables except GDP growth rates, we make projections of their future values based on trend analysis. GDP growth rates are developed on the basis of predictions by other research agencies.

By applying the projected values of explanatory variables to the estimated labour supply function, a baseline prediction of labour supply is generated (Table 12.5). In order to ascertain the effect of raising the female labour force participation rate on labour supply, we examine two situations in which the participation rate is increased by 0.1% and 0.2%, respectively, each quarter. The new prediction less baseline prediction as a percent of baseline prediction is taken as a measure of the effect of FLFPR on labour supply. Over the 5-year period, on average, an increase of 0.2% in FLFPR results in an increase of 0.090% to 0.095% in labour supply.

Table 12.1 Logit estimate results of female labour force participation, sample 1 (sample size = 25,949 females), October 1998

Explanatory variable	*Coefficient*	*t ratio*
Constant	-1.2392*	-5.4066
Urbanization		
Rural area	----	----
City or metropolitan area	-0.2650*	-10.6417
Age		
16 – 25	----	----
26 – 35	0.2295*	4.2658
36 – 45	0.2385*	3.9356
46 – 55	0.1531*	3.3719
56 – 65	0.0884*	4.3762
Years of schooling		
0 – 6	----	----
7 – 12	0.5708*	16.7390
13 and over	0.6231*	11.8686
Work Experience		
No	----	----
Yes	0.1448***	1.7616
Marital status		
Married spouse deceased	----	----
Married, spouse present	0.4352*	6.4955
Not married	0.7631*	9.1499
Divorced	0.1187*	2.9236

Notes: ---- indicates the base or omitted category

* indicates significant at 1% level

*** indicates significant at 10% level

Source: *Labour Force Survey, October 1998*, Directorate – General of Budget, Accounting and Statistics, Taiwan.

Table 12.2 Logit estimate results of female labour force participation, sample 2 (sample size = 14,100 females), 1997

Explanatory variable	*Coefficient*	*t ratio*
Constant	-1.0173*	-8.0039
Urbanization		
Rural area	- - - -	- - - -
City or metropolitan area	-0.5793*	-10.6489
Town	-0.3147*	-5.3703
Age		
16 – 25	- - - -	- - - -
26 – 35	0.2829*	3.6598
36 – 45	0.5374*	6.3075
46 – 55	0.1685*	5.1061
56 – 64	0.0834*	8.0192
Years of schooling		
0 – 6	- - - -	- - - -
7 – 9	0.3515*	5.6330
10 - 12	0.5741*	9.9154
13 and over	0.6425*	9.8092
Marital status		
Married spouse deceased	- - - -	- - - -
Married, spouse present	0.6519*	5.6687
Not married	0.7826*	4.8730
Divorced	0.3352*	3.8485

Notes: - - - - indicates the base or omitted category

* indicates significant at 1% level

Source: *Household Income and Expenditure Survey, 1997,* Directorate – General of Budget, Accounting and Statistics, Taiwan.

Table 12.3 Logit estimate results of female labour force participation, sample 3 (sample size = 6,972 households, married, spouse present), 1997

Explanatory variable	*Coefficient*	*t ratio*
Constant	-1.1234*	-7.1691
Urbanization		
Rural area	- - - -	- - - -
City or metropolitan area	-0.9558*	-11.8881
Town	-0.5061*	-6.0178
Number of Children	0.0746*	2.9255
Household income (Average=NT$ 1,197,983)	0.0070*	14.0173
Age(Average=40.62years)		
16 – 25	- - - -	- - - -
26 – 35	0.5597*	4.1613
36 – 45	0.7336*	5.3508
46 – 55	0.0707*	2.8740
56 – 65	-0.4354*	-4.1427
Years of schooling		
0 – 6	- - - -	- - - -
7 – 9	0.0356**	2.0136
10 – 12	0.3969*	5.3780
13 and over	0.5992*	6.0161
Same household as parents-in-law		
No	- - - -	- - - -
Yes	0.1354**	2.0609

Notes: - - - - indicates the base or omitted category
* indicates significant at 1% level
** indicates significant at 10% level

Source: *Household Income and Expenditure Survey, 1997*, Directorate – General of Budget, Accounting and Statistics, Taiwan.

Table 12.4 Predictions of female labour participation

Urbanization	*Age*	*Education*	*Working Experience*	*Marital Status*	*Participation Rate (%)*
Rural	16-25	Elementary School	No	Not Married	38.3173
Rural	26-35	Elementary School	No	Not Married	43.8661
Rural	36-45	Elementary School	No	Not Married	44.0878
Rural	46-55	Elementary School	No	Not Married	41.9941
Rural	56-64	Elementary School	No	Not Married	40.4271
Rural	16-25	High School	No	Married	44.1963
Rural	26-35	High School	No	Married	49.9075
Rural	36-45	High School	No	Married	50.1325
Rural	46-55	High School	No	Married	47.9982
Rural	56-64	High School	No	Married	46.3863
City	16-25	High School	Yes	Not Married	49.3625
City	26-35	High School	Yes	Not Married	55.0824
City	36-45	High School	Yes	Not Married	55.3050
City	46-55	High School	Yes	Not Married	53.1853
City	56-64	High School	Yes	Not Married	51.5720
City	16-25	College	Yes	Not Married	50.6700
City	26-35	College	Yes	Not Married	56.3727
City	36-45	College	Yes	Not Married	56.5939
City	46-55	College	Yes	Not Married	54.4850
City	56-64	College	Yes	Not Married	52.8768
City	16-25	College	Yes	Married	42.5289
City	26-35	College	Yes	Married	48.2108
City	36-45	College	Yes	Married	48.4355
City	46-55	College	Yes	Married	46.3063
City	56-64	College	Yes	Married	44.7025

Table 12.5 Effects of raising female labour force participation rate on labour supply, First Quarter 1999 to Fourth Quarter 2003

Prediction / Time	❶ Baseline prediction of labour supply	❷ New Prediction of labour supply FLFPR raised by 0.1%	❸ New prediction of labour supply FLFPR raised by 0.2%	❹ =(❷-❶)÷❶×100 (%)	❺ =(❸-❶)÷❶×100 (%)
1st quarter 1999	9,610,142	9,614,721	9,619,301	0.047652	0.095303
2nd quarter 1999	9,638,279	9,642,858	9,647,437	0.047513	0.095025
3rd quarter 1999	9,668,905	9,673,484	9,678,064	0.047362	0.094724
4th quarter 1999	9,697,906	9,702,485	9,707,064	0.047221	0.094441
1st quarter 2000	9,725,363	9,729,942	9,734,521	0.047087	0.094174
2nd quarter 2000	9,753,274	9,757,854	9,762,433	0.046952	0.093905
3rd quarter 2000	9,780,908	9,785,487	9,790,066	0.046820	0.093640
4th quarter 2000	9,808,141	9,812,720	9,817,300	0.046690	0.093380
1st quarter 2001	9,835,588	9,840,167	9,844,747	0.046559	0.093119
2nd quarter 2001	9,862,596	9,867,175	9,871,755	0.046432	0.092864
3rd quarter 2001	9,889,615	9,894,194	9,898,774	0.046305	0.092610
4th quarter 2001	9,916,401	9,920,980	9,925,559	0.046180	0.092360
1st quarter 2002	9,943,199	9,947,778	9,952,358	0.046056	0.092111
2nd quarter 2002	9,969,766	9,974,345	9,978,925	0.045933	0.091866
3rd quarter 2002	9,996,306	10,000,885	10,005,465	0.045811	0.091622
4th quarter 2002	10,022,738	10,027,318	10,031,897	0.045690	0.091380
1st quarter 2003	10,049,064	10,053,644	10,058,223	0.045570	0.091141
2nd quarter 2003	10,075,571	10,080,150	10,084,729	0.045451	0.090901
3rd quarter 2003	10,101,564	10,106,143	10,110,722	0.045334	0.090667
4th quarter 2003	10,127,452	10,132,032	10,136,611	0.045218	0.090435

Factor Analysis of Changes in Labour Force Participation

The labour force participation rate depends on the number of people aged 15 years and over and the willingness of people in this age group to participate in the labour market. Analysing changes in labour force participation by demographic characteristics does not reveal whether the changes are due to demographic characteristic changes alone or to a shift in willingness to participate in the labour market. In this section, attention is focused on decomposition of changes in the labour force participation rate into components attributable to a change in population structure and to the effect of a shift in willingness to participate.

If we divide labour force (*LF*) into four groups corresponding to ages 16-24, 25-49, 50-64, and 65 and over, then total labour force participation rate (*LFPR*) can be decomposed as follows:

$$LFPR = \frac{LF}{P} = \frac{LF_1}{P} + \frac{LF_2}{P} + \frac{LF_3}{P} + \frac{LF_4}{P}$$

$$= \left(\frac{P_1}{P}\cdot\frac{LF_1}{P_1}\right) + \left(\frac{P_2}{P}\cdot\frac{LF_2}{P_2}\right) + \left(\frac{P_3}{P}\cdot\frac{LF_3}{P_3}\right) + \left(\frac{P_4}{P}\cdot\frac{LF_4}{P_4}\right)$$

$$= \sum_{i=1}^{4}\left(\frac{P_i}{P}\cdot\frac{LF_i}{P_i}\right) = \sum_{i=1}^{4} PR_i \cdot LFPR_i \quad \text{......... (9)}$$

where P is the population aged 15 years and over, LF_i is the labour force of the ith age group, P_i is the population of the ith age group, $LFPR_i$ is the labour force participation rate of the ith age group, and PR_i is the population structure of the ith age group.

In terms of changes in $LFPR_i$ over time, equation (9) can be expressed as

$$\Delta LFPR_t = \sum_{i=1}^{4} \Delta LFPR_{i,t} \bullet PR_{i,t-1} + \sum_{i=1}^{4} \Delta PR_{i,t} \bullet LFPR_{i,t-1} + \sum_{i=1}^{4} \Delta LFPR_{i,t} \bullet \Delta PR_{i,t} \quad \text{......... (10)}$$

where

$$\Delta LFPR_t = LFPR_t - LFPR_{t-1}$$

$$\Delta LFPR_{i,t} = LFPR_{i,t} - LFPR_{i,t-1}$$

$$\Delta PR_{i,t} = PR_{i,t} - PR_{i,t-1}$$

Equation (10) shows that changes in LFPR over time are attributable to three separate effects. The first term on the right-hand side of the equation,

$$\sum_{i=1}^{4} \Delta LFPR_{i,t} \cdot PR_{i,t-1}$$

shows the effect of changes in $LFPR_i$ on D*LFPR*, while the population structure remains unchanged. The second term,

$$\sum_{i=1}^{4} \Delta PR_{i,t} \cdot LFPR_{i,t-1}$$

shows the effect of changes in the population structure on $\Delta LFPR$, while $LFPR_i$ remains unchanged. The third term,

$$\sum_{i=1}^{4} \Delta LFPR_{i,t} \cdot \Delta PR_{i,t}$$

shows the cross effect of changes in both $LFPR_i$ and PR_i on $\Delta LFPR$. The cross effect is normally very small and can be ignored.

Referring to Table 12.6, in the age group 15-24, changes in both the population structure of this age group and in its willingness to participate in the labour market have had negative effects on changes in total female labour force participation rate. Over long periods of time, due to a shift in demographic structure, the proportion of this age group to total population has declined steadily, resulting in a negative effect on total female labour force participation. Opportunities for higher education in Taiwan have improved continuously due to the increase in the number of colleges and the rise in college enrolment. Many females in the age group 15-24 choose to pursue a college education and delay entry into the labour market, which is another reason for the negative effect on the willingness of this age group to participate in the labour market.

The age group 25-49 is the core labour force group generating the most significant effect on changes in total labour force participation rate. With a few exceptions, changes in both the willingness to participate in the labour market and population structure of this age group exert pronounced positive effects on changes in the total female labour force participation rate over the period 1980 to 1997. This age group contributes the most to the increase in labour force participation.

In the age group 50-64, although changes in willingness to participate show positive effects most of the time over the period 1980 to 1997, the magnitude of this effect is less than that of the 25-49 age group. Changes in population structure exhibit minor positive effects from 1980 to 1992, but negative effects since 1993. The negative effect of changes in population structure in the later period is due to an increase in early retirement, contributing to a minor decrease in female labour force participation.

Table 12.6 Factor analysis of changes in female labour force participation rates by age group, Taiwan, 1980 – 1997

	Age 15 – 24		Age 25 – 49		Age 50 – 64		Age 65 and over		
Year	*Effects of changes in LFPR*	*Effects of changes in population structure*	*Effects of changes in LFPR*	*Effects of changes in population structure*	*Effects of changes in LFPR*	*Effects of changes in population structure*	*Effects of changes in LFPR*	*Effects of changes in population structure*	*Changes in total female labour force participation rate*
1980	-0.20424	-0.32803	0.3462	0.18469	0.02961	0.03233	-0.01449	0.00175	0.02800
1981	-0.20484	-0.37804	-0.13563	0.18309	0.00895	0.03131	0.00538	0.00251	-0.49300
1982	-0.22504	-0.31934	0.71784	0.24843	0.10579	0.02252	0.01004	-0.00052	0.53690
1983	0.29809	-0.32527	2.07858	0.30484	0.43721	0.01651	0.04629	0.00039	2.81710
1984	-0.06913	-0.39205	1.1296	0.25528	0.2316	0.0264	0.00689	0.00516	1.18390
1985	-0.26577	-0.4006	0.44898	0.25098	0.07177	0.03825	0.03986	0.00756	0.16000
1986	0.34986	-0.41273	1.57868	0.25189	0.19008	0.03937	0.03891	0.00870	2.04900
1987	-0.00884	-0.42631	0.96821	0.31554	0.16149	0.03863	-0.00075	0.00700	1.03500
1988	-0.51135	-0.44321	-0.25348	0.27488	-0.07468	0.04907	-0.03480	0.00726	-0.98700
1989	-0.23651	-0.39765	0.09199	0.28938	-0.03928	0.04004	0.03876	0.00690	-0.20800
1990	-0.59675	-0.30012	-0.21342	0.2592	-0.02597	0.03091	0.00805	0.00580	-0.85400
1991	-0.29887	-0.28731	0.25899	0.09024	0.08899	0.0294	0.00830	0.01018	-0.10190
1992	-0.38507	-0.2907	0.88906	0.1227	0.07549	0.02021	-0.00427	0.00950	0.43800
1993	-0.271	-0.19163	0.52882	-0.02975	0.03357	-0.00098	-0.00614	0.00959	0.06400
1994	0.08883	-0.26224	0.60951	0.07522	0.00369	-0.02505	0.00360	0.00863	0.50700
1995	-0.27554	-0.2393	0.32326	0.15068	0.07079	-0.09627	0.01844	0.00978	-0.05800
1996	-0.14284	-0.13412	0.68711	0.1085	0.00997	-0.10722	-0.02453	0.00853	0.41900
1997	-0.20543	-0.01722	0.19697	-0.0378	-0.01715	-0.03178	-0.01156	0.00765	-0.11900

Source: Aremos Economic Statistics Data Bank

The 65 and over age group consists mainly of retirees. Changes in the willingness to participate in the labour market and in the population structure of this age group have negligible effects on changes in total labour force participation. This is consistent with the fact that retirees are out of the labour market so that changes in the willingness to work and population structure of this age group should play no role in affecting total labour force participation.

Conclusions

The female labour force participation rate in Taiwan is relatively low compared with that of industrialized countries. Female labour is more flexible than male labour in its entry to and exit from the labour market. The abundant supply of working-age females is a potential source of increased labour supply that can be properly channelled to meet the rising demand for labour by industries.

This study applies logit analysis to factors associated with labour force participation. Education is found to play the most important role in raising participation. Work experience has a positive effect on participation, whereas urbanization has a negative effect. Age has a positive effect on participation for females at the early stage of their life, but its effect diminishes after middle age. Women not married have higher participation than married women. Residing in the same household as parents-in-law helps to raise participation. Household income has a small positive effect on participation for females. The probability of participation in the labour market is highest when the woman is not married, has working experience, and has a university degree.

Based on an estimated labour supply function, it is estimated that, over the 5-year period 1999 to 2003, an increase in female labour force participation rate of 0.2% results in an increase of 0.09% to 0.095% in total labour supply.

Changes in labour force participation rate can be decomposed into components due to change in population structure and to the effect of a shift in willingness to participate. In the age group 15-24, both changes in the population structure of this group and in its willingness to participate in the labour market have a negative effect on changes in the total female labour force participation rate. The 25-49 age group generates the most significant positive effect on changes in the total female labour force participation rate. In the age group 50-64, a change in willingness to participate has a positive effect but is less significant than that of the age group 25-49. The effect of changes in the population structure of this group was positive from 1980 to 1992 and negative since 1993.

ENDNOTES

1 As shown in elementary probability theory, $\hat{P}_i$, the proportion of successes (here joining the labour force), follows the binomial distribution with mean equal to true P_i and variance equal to

$$P_i(1-P_i)/N_i$$

and that as N_i increases indefinitely the binomial distribution approximates the normal distribution. For details, see Henry Theil (1970).

REFERENCES

Barrow, L. (1996). An analysis of women's labour force participation following first birth. Princeton University, *Industrial Relations Section*, Working Paper: 363, June.

Bauer, J. (1990). Demographic change and Asian labour markets in the 1990s. *Population and Development Review*, 614, 615-645.

Becker, G. S. (1965). A theory of the allocation of time. *Economic Journal*, 75, 493-517.

Brooks, R. (1991). Male and female labour force participation in New Zealand 1965-1990: A cointegration analysis. *New Zealand Economic Papers*, 25(2), 219-251.

Chang, C. (1978). A review of female labour force participation. *Economic Essays*, 8, 275-284.

Eissa, N., and Liebman, J. B. (1995). Labour supply responses to the earned income tax credit, *National Bureau of Economic Research* Working Paper: 5158, June.

Fernandez, A. I., and Rodriguez-Poo, J. M. (1997). Estimation and specification testing in female labour participation models: Parametric and semiparametric methods. *Econometric Reviews*, 16(2), 229-247.

Friedman, M. (1957). *A theory of the consumption function*. Princeton, NJ: Princeton University Press.

Goldin, C. (1994). The U-shaped female labour force function in economic development and economic history, *National Bureau of Economic Research*, Working Paper: 4707.

Granger, C. W. J. (1980). Testing for causality: A personal viewpoint, *Journal of Economic Dynamics and Control*, 1, 321-346.

Humphrey, J. (1996). Responses to recession and restructuring: Employment trends in the Sao Paulo metropolitan region, 1979-87, *Journal of Development Studies*, 33(1), 40-62.

Kao, Y., and Chen, S. (1994). Causality analysis of female labour participation behavior in Taiwan, *Journal of Women and Both Sexes*, 5, National Taiwan University Center for Population Studies (in Chinese).

Lo, C. (1986). Reconsideration of labour participation of married women, *Academia Economic Papers*, 14(1), 113-130.

Mincer, J. (1962). Labour force participation of married women. In H. G. Lewis (Ed.), *Aspects of labour economics* (pp. 63-105). Princeton, NJ: Princeton University Press.

Mincer, J. (1966). Labour force participation and unemployment: A review of recent evidence. In R. A. Gordon and M. S. Gordon (Eds.), *Prosperity and unemployment* (pp. 73-112). New York: Wiley.

Mincer, J. (1985). Intercountry comparisons of labour force trends and of related development: An overview, *Journal of Labour Economics*, 3(1)s, s1-s32.

Ribar, D. C. (1995). A structural model of child care and labour supply of married women, *Journal of Labour Economics*, 13(3), 558-597.

Theil, H. (1970). On the relationships involving qualitative variables, *American Journal of Sociology*, 76, 103-154.

Unni, J. (1993). Labour supply decisions of married women in rural India, *Yale Economic Growth Center Discussion Paper* 691.

13

Fractional Cointegration Analysis of Purchasing Power Parity Between Taiwan and Major Industrial Countries

Chingnun Lee

Associate Professor, Graduate Institute of Economics, National Sun Yat-Sen University

Introduction

The doctrine of purchasing power parity (PPP) is an important element of international macroeconomics. It means that the nominal exchange rate should be equal to the ratio of prices in two countries. Or simply, when measured in the same units, the currencies of different countries should command the same basket of goods. Otherwise, international arbitrage should bring adjustments in prices, exchange rates, or both, which ultimately restore parity. What that implies is an equilibrium relationship between price and the exchange rate in any country.

Although many models imply PPP, the empirical findings are mixed as to whether PPP is obtained in practice. Many empirical studies report significant deviations from PPP in the short run. Moreover, the validity of PPP in the long run remains controversial. For example, Roll (1979), Frenkel (1981), Adler and Lehmann (1983), Hakkio (1986), and Mark (1990) found that real exchange rates closely followed a random walk, suggesting that shocks have infinitely long-lived effects. On the other hand, based on a cointegration test, Kim (1990) and Wu (1995) find support for PPP.

While most studies to date allow for a long-run cointegration process between the nominal exchange rate and relative prices to be manifested over time, there is a need to model appropriately the low frequency dynamics inherent in the temporal characteristics of a number of economic variables. Based on fractional difference analysis, Diebold, Husted, and Rush (1991) provided evidence of mean reversion in the real exchange rate under the gold standard.

The test for cointegration often presumes the order of integration of the equilibrium error to be an integer (always equal to zero) (e.g., Engel and Granger, 1987; Johansen, 1991). A system of economic variables, however, can be fractionally cointegrated such that the equilibrium error follows a fractionally integrated process of the kind in Granger and Joyeux (1980) and Hosking (1981).

In this study, a generalized notion of cointegration called fractional cointegration is used to re-examine the long-run PPP hypothesis in respect of Taiwan with other major industrial countries. The chapter is mainly concerned with the case in which the nominal exchange rate, and domestic and foreign price indices are all I (1).[1] If the equilibrium error can be found to be I (1-b) with b greater than 0, then the exchange rate and relative prices are fractionally cointegrated, and the effect of a shock to the system will eventually die out.[2] This means that the equilibrium error in PPP is mean reverting, and the theory is verified for Taiwan.

THE PPP FORMULATION

The notion of PPP requires the equilibrium exchange rate to reflect the relative purchasing power of two currencies. Assume there are *m* goods in the world, then we can define the domestic and foreign price index to be

$$P_t = \sum_{i=1}^{m} \omega_{it} P_{it} \text{ , } 0 < \omega_{it} < 1 \text{ , } \sum_{i=1}^{m} \omega_{it} = 1 \text{ , and}$$

$$P_t^* = \sum_{i=1}^{m} \omega_{it}^* P_{it}^* \text{ , } 0 < \omega_{it}^* < 1 \text{ , } \sum_{i=1}^{m} \omega_{it}^* = 1,$$

where P_{it}, P_{it}^* are the price of the i-th item of domestic and foreign good in time period t respectively, and ω_{it} ,ω_{it}^* are the weights of the i-th goods in the construction respectively of the domestic price index and the foreign price index in the same period of time.

According to the law of one price, in the absence of transportation cost, the results of free trade will bring the price of all traded goods to equality implying that the equilibrium exchange rate S_t

$$S_t = \frac{P_t}{P_t^*}$$

Empirically, we use the model

$$s_t = c + \beta_1 p_1 + \beta_2 p_t^* + v_t \text{ ,}$$

where s_t is the nominal spot exchange rate, p_t is the domestic aggregate price level, and p_t^* the foreign aggregate price level, all in natural logarithms. Under the hypothesis of I (1) in all three variables (s_t, p_t, and p_t^*), a necessary condition for the establishment of PPP is the mean-reverting behaviour in the error v_t.[3] In this chapter, our econometric method is designed to estimate the parameter β's and to test for mean reversion v_t.[4]

Econometrics Methodology

Cointegration and Fractional Cointegration

Following the warning of spurious regression by Granger and Newbold (1974) and Phillips (1986), econometricians have been more cautious about dealing with non-stationary data in time-series analysis. Nelson and Plossor (1982), using the Dickey-Fuller unit root test, show that most macroeconomic data in the US are non-stationary. Subsequent studies have been concerned with the power, size distortion, and possible structural breaks of the unit root tests (Phillips and Perron, 1988; Gil-Alana and Robison, 1997). Engel and Granger (1987) define the concept of cointegration as follows. Among a set of non-stationary data, a set of (n×1) variables z_t (in our case, $z_t=(s_t, p_t, p_t^*)'$) are all I(d) if we can find a (n×r) matrix α such that $\alpha' z_t \sim I(d-b)$, with b>0; we then call the matrix α the cointegrating matrix and the I(d-b) process $\alpha' z_t$ is defined as equilibrium error. In recent times, an enormous number of studies have examined the case where d and b are all equal to 1; they estimate α and test whether $\alpha' z_t$, the equilibrium error, is I(0) or not.

Engel and Granger's two step method used the process of ordinary least squares (OLS) to estimate α and applied the unit root test in the investigation of the OLS residual to decide whether z_t is cointegrated or not. Johansen (1991) used a Gaussian VAR model to estimate α and to overcome the drawbacks in Engel and Granger's two step method that estimated only one cointegration vector.

However, these two standard tests of cointegration are too restrictive to permit acceptance or rejection of PPP as a long run condition. According to Granger (1986), there is no requirement for the error term to be I(0). The economic intuition behind this condition makes it obvious that, unless v_t reverts back to its mean, a shock to z_t will witness it deviating from equilibrium in a permanent fashion. The statistical properties of v_t being fractionally integrated, after the fashion of Granger and Joyeux (1980) and Hosking (1981), are sufficient to restore equilibrium.[5] That is, $v_t \sim I(d-b)<1$, a long memory process, would imply mean-reversion. Therefore, all of the requirements for a set of I(1) variables to be fractionally cointegrated rest on the estimation of the integration parameter in the error term. See, for example, Cheung and Lai (1993).

Estimation of Fractionally Cointegrated Relationships

A combination of Engel-Granger two-step procedures and the method of estimating a fractionally integrated process are used to perform the empirical results. The approach is summarized in the following steps:

- Determine the univariate properties of each of the time series involved with respect to integration via a battery of unit root tests.

- Given that the variables share the common integration processes, conduct OLS regressions and compute the OLS residuals (v_t)s or the equilibrium errors.
- Examine whether the v_ts are I(1-b) with b>0 using an appropriate estimation technique designed to detect fractionally integrated processes. Use the conditional sum of squares (CSS)[6] method (Chung and Baillie (1993)) to estimate the fractionally integrated process in procedure (iii)

We now consider estimating the p+q+3 dimensional vector of parameters $\lambda = (\mu\ \beta)$, where $\beta = (d, \phi_1, \phi_2, \cdots, \phi_p, \theta_1, \theta_2, \cdots, \theta_q, \sigma_2)$ in the ARFIMA (p, d, q) process $\phi(L)(1-L)^d(y_t - \mu) = \theta(L)\varepsilon_t$.

Under the assumption of normality in the innovation, Sowell (1992) was able to derive the full maximum likelihood estimator (MLE) for the ARFIMA (p, d, q) process. The logarithm of the likelihood in the time domain can be expressed as

$$L(\lambda) = -\frac{T}{2}\log 2\pi - \frac{1}{2}\log|\Sigma| - \frac{1}{2}\left[(y-\mu)'\,\Sigma^{-1}(y-\mu)\right]$$

where y is the T dimensional vector of y_t and Σ is the corresponding T×T autocovariance matrix.

Chung and Baillie (1993) considered an alternative conditional sum of squares estimator which minimizes

$$S(\lambda) = \frac{1}{2}\log\sigma^2 + \frac{1}{2\sigma^2}\sum_{t=1}^{T}\varepsilon_t^2$$
$$= \frac{1}{2}\log\sigma^2 + \frac{1}{2\sigma^2}\sum_{t=1}^{T}\left[\phi(L)\theta(L)^{-1}(1-L)^d(y_t-\mu)\right]^2$$

If the initial observations of $y_0, y_{-1}, y_{-2}, ..$ are assumed fixed, then minimizing the CSS function will be asymptotically equivalent to MLE.

Cheung and Lai (1993) have shown that OLS in procedure (ii) is a consistent estimator for fractional cointegration parameters[7] that would inspire the use of OLS residuals to act as equilibrium errors.

Empirical Results

Data Description and Source

We use monthly data for the spot exchange rate (s_t) between Taiwan and the US, Canada, France, the UK, and Germany. The consumer price index (CPI) in each country stands for the price level (p_t for Taiwan and p_t^* for the foreign country). All the data are obtained from the database of AREMOS /UNIX in the Ministry of Education in Taiwan. Since from the beginning of 1980 the exchange rate in Taiwan has been less regulated, the sample periods in the study are from January 1980 to the end of 1995.

Unit roots test

The following tables show the results of the Augmented Dickey-Fuller test (1979) and the Phillips and Perron test (1988). The results indicate that the unit roots hypothesis cannot be rejected on the spot exchange rate and the CPI.

Table 13.1 Unit root tests on exchange rate (level)

Test statistics	*AR*	*Twn/US*	*Twn/UK*	*Twn/Ger*	*Twn/Fra*	*Twn/Can*
τ_μ	4	-0.760	-2.847	-3.420*	-4.379*	-0.921
$r_\mu=T(\rho_\mu-1)$	4	-0.556	-6.172	-16.29*	-11.67	-1.197
τ_τ	4	-1.416	-2.772	-2.714	-3.731*	-1.738
$r_\tau=T(\rho_\tau-1)$	4	-1.990	-12.070	-17.921	-11.423	-6.784
$\tau_\mu\bigstar$	4	-0.574	-2.084	-2.714	-2.927	-0.677
$r_\mu\bigstar$	4	-0.677	-4.811	-13.915	-8.179	-0.897
$\tau_\tau\bigstar$	4	-0.971	-1.974	-2.756	-2.301	-1.866
$r_\tau\bigstar$	4	-1.905	-8.834	-13.813	-7.432	-7.220

* means significance in Schwert (1987).

τ_μ, r_μ, τ_τ and r_τ represent the test statistics of the Augmented Dickey-Fuller test with drift and with drift and time trend respectively. $\tau_\mu\bigstar$, $r_\mu\bigstar$, $\tau_\tau\bigstar$, $r_\tau\bigstar$ represent the test statistics of the Phillips and Perron (1988) test with the same cases.

Table 13.2 Unit root tests on CPI (level)

Test statistics	*AR*	*US*	*UK*	*Ger*	*Fra*	*Can*	*Twn*
τ_μ	4	-0.982	-1.060	-0.309	-5.925*	-3.385*	-0.492
$r_\mu=T(\rho_\mu-1)$	4	-0.174	-0.312	-0.102	-1.442	-0.971	-0.565
τ_τ	4	-1.840	-0.982	-1.241	-3.774	-1.538	-1.404
$r_\tau=T(\rho_\tau-1)$	4	-4.172	-2.596	-1.848	-3.000	-2.469	-5.111
$\tau_\mu\bigstar$	4	-1.450	-1.307	-0.403	-9.320	-4.743	-0.969
$r_\mu\bigstar$	4	-0.359	-0.432	-0.182	-1.767	-1.310	-0.959
$\tau_\tau\bigstar$	4	-1.767	-1.171	-0.827	-3.623	-1.361	-2.085
$r_\tau\bigstar$	4	-5.181	-1.905	-1.487	-3.062	-2.329	-6.980

* means significance in Schwert (1987).

For avoiding the second unit root, the same tests are applied to the first differences of each variables. The results indicate that the variables considered are integrated of order less than 2.

Table 13.3 Unit root tests on exchange rate (first differenced)

Test statistics	*AR*	*Twn/US*	*Twn/UK*	*Twn/Ger*	*Twn/Fra*	*Twn/Can*
τ_μ	4	-4.222*	-6.684*	-7.108*	-6.512*	-6.656*
r_μ	4	-78.174*	-177.552*	-186.83*	-166.53*	-188.54*
τ_τ	4	-9.769*	-13.250*	-13.468*	-13.248*	-13.662*
r_τ	4	-139.21*	-192.55*	-195.57*	-193.98*	-194.67*

* means significance in Schwert (1987).

Table 13.4 Unit root tests on CPI (first differenced)

Test statistics	*AR*	*US*	*UK*	*Ger*	*Fra*	*Can*	*Twn*
τ_μ	4	-5.936*	-7.122*	-5.481*	-3.981*	-4.329*	-7.344*
r_μ	4	-120.98*	-178.59*	-117.23*	-61.931*	-94.471*	-239.92*
τ_τ	4	-9.184*	-11.717*	-10.498*	-7.956*	-13.210*	-14.449*
r_τ	4	-115.86*	-150.61*	-149.23*	-96.776*	-227.34*	-184.76*

* means significance in Schwert (1987).

Johansen's Cointegration Tests

For comparison with our main results on fractional cointegration, we report the results of Johansen (1991) tests for cointegration. Aside from Taiwan and England, the Trace test[8] and the maximum eigenvalue test[9] both indicate that there is one cointegration relationship between Taiwan and other countries.

Table 13.5 Johansen tests for cointegration

	Trace Test			*Maximum Eigenvalue Test*		
	$\gamma = 0$	$\gamma \leq 1$	$\gamma \leq 2$	$\gamma = 0$	$\gamma \leq 1$	$\gamma \leq 2$
US	48.945020*	15.793907	4.054541	33.151113*	11.739265	4.054641
UK	34.153320*	16.987837	1.768481	17.165495	15.230989	1.756848
Ger	44.223199*	13.026990	0.506974	31.196209*	12.520017	0.506972
Fra	65.279971*	19.835543	0.106604	45.444428*	19.728938	0.106604
Can	48.945100*	16.759070	2.840744	32.186110*	13.918333	2.840744

However, Johansen tests are based on the assumption of the equilibrium error being a dichotomy on I (0) and I (1). We relax this strong assumption and use a fractional cointegration framework that allows the equilibrium to be any real number between 0 and 1. We have the following results.

Fractional Cointegration Tests

1) *Taiwan and US case*

We first use OLS to get the results:

$$v_t = s_t - 5.4505299 - 1.0975400\, p_t + 1.5502365\, p_t^*$$

Then we estimate v_t in the models of ARFIMA (p,d,q)[10] by CSS, and get the following.

Table 13.6 CSS estimation of Taiwan/US fractional cointegration

$$C(L)(1-L)^b v_t = D(L)\mu_t$$

ARFIMA (3 , 0.45162 , 1)

The Main Estimation Results

Fractional Differencing with Estimated Fraction: b=0.45162

Parameter	*Estimate*	*S.E.*	*T-Ratio*
FRAC.D.	0.45162	0.24443	1.84762
AR(1)	1.11758	0.16606	6.72995
AR(2)	-0.36143	0.13567	-2.66399
AR(3)	0.18447	0.07712	2.39193
OMEGA	0.00029	0.00003	10.09950

Value of the Likelihood Function	-540.373536
Number of Observations	204
Residual Diagnostics:	
Skewness Measure	-1.93929
Kurtosis Measure	13.97999
Mean	0.00016
Variance	0.00029

Ljung-Box Statistic:

Q(10)=7.83882	Q(15)=12.01071
Q(20)=19.20534	Q(25)=19.85728

The identical procedures are applied to the case of Taiwan and the UK, France, Germany, and Canada. The results are as follows.

2) Taiwan and UK case

Table 13.7 CSS estimation of Taiwan/UK fractional cointegration

$$C(L)(1-L)^b v_t = D(L)\mu_t$$

ARFIMA (1 , 0.19608 , 0)

The Main Estimation Results

Fractional Differencing with Estimated Fraction: b=0.19608

Parameter	*Estimate*	*S.E.*	*T-Ratio*
FRAC.D.	0.19608	0.06392	15.64485
AR	0.84919	0.07134	11.90272
OMEGA	0.00123	0.00012	10.09951

Value of the Likelihood Function	-394.15608
Number of Observations	204
Residual Diagnostics:	
Skewness Measure	0.09295
Kurtosis Measure	3.85461
Mean	0.00065
Variance	0.00123
Ljung-Box Statistic:	
Q(10)=7.97232	Q(15)=15.96899
Q(20)=19.17935	Q(25)=21.56871

where $v_t = s_t - 6.5343415 - 0.38846224\, p_t + 0.97946783\, p_t^*$

3) Taiwan and France case

Table 13.8 CSS estimation of Taiwan/Fra fractional cointegration

$$C(L)(1-L)^b v_t = D(L)\mu_t$$

ARFIMA (1 , 0.08684 , 1)

The Main Estimation Results
Fractional Differencing with Estimated Fraction: b=0.08684

Parameter	*Estimate*	*S.E.*	*T-Ratio*
FRAC.D.	0.08684	0.17945	0.48391
AR	0.90746	0.05379	16.87162
MA	-0.01210	0.16401	-0.07379
OMEGA	0.00150	0.00015	10.09950

Value of the
Likelihood Function -373.501937

Number of Observations 204

Residual Diagnostics:

Skewness Measure	-1.30566
Kurtosis Measure	12.46038
Mean	0.00031
Variance	0.00150

Ljung-Box Statistic:

Q(10)=3.86222	Q(15)=10.88607
Q(20)=17.79189	Q(25)=18.72823

where $v_t = s_t - 2.4633965 - 0.92942325\, p_t + 1.1224607\, p_t^*$

4) Taiwan and Germany case

Table 13.9 CSS estimation of Taiwan/Ger fractional cointegration

$$C(L)(1-L)^b v_t = D(L)\mu_t$$

ARFIMA (1 , 0.06737 , 0)

The Main Estimation Results
Fractional Differencing with Estimated Fraction: b=0.06737

Parameter	*Estimate*	*S.E.*	*T-Ratio*
FRAC.D.	0.06737	0.08463	0.79602
AR	0.91167	0.04239	21.50516
OMEGA	0.00148	0.00015	10.09950

Value of the Likelihood Function	-375.185137
Number of Observations	204
Residual Diagnostics:	
Skewness Measure	1.07103
Kurtosis Measure	10.57739
Mean	0.00028
Variance	0.00148

Ljung -Box Statistic:

Q (10)=5.26374	Q (15)=12.61493
Q (20)=17.49218	Q (25)=18.83048

where $v_t = s_t - 2.5661275 - 0.13657150\, p_t + 0.086967854\, p_t^*$

5) Taiwan and Canada case

Table 13.10 CSS estimation of Taiwan/Can fractional cointegration

$$C(L)(1-L)^b v_t = D(L)\mu_t$$

ARFIMA (1 , 0.13953 , 0)

The Main Estimation Results
Fractional Differencing with Estimated Fraction: b=0.13953

Parameter	*Estimate*	*S.E.*	*T-Ratio*
FRAC.D.	0.13953	0.16977	0.82187
AR	0.93734	0.04053	23.12918
MA	-0.11330	0.16208	-0.69906
OMEGA	0.00047	0.00005	10.09950

Value of the
Likelihood Function -491.603366

Number of Observations 204

Residual Diagnostics:

Skewness Measure -3.15739

Kurtosis Measure 30.22404

Mean -0.00001

Variance 0.00047

Ljung-Box Statistic:

Q (10)=1.80858 Q (15)=5.58592

Q (20)=12.40034 Q (25)=13.21329

where $v_t = s_t - 7.0470120 - 0.12012301\, p_t + 0.73489220\, p_t^*$

All the results support the existence of PPP between Taiwan and other major industrial countries. This conclusion is consistent with the work of Wu (1995).

CONCLUSION

The main idea of PPP rests on the logic that nominal exchange rates can be determined by relative purchasing power between two currencies. Relaxing the strict assumption on the equilibrium error of Johansen (1991), we empirically test for fractional cointegration between Taiwan and five of its major trading partners during the period of exchange rate deregulation since 1980.

The results support the operation of the PPP theory in Taiwan with these countries. Taiwan has equilibrium errors of PPP that are found to be long memory processes when it is tied with the US and the UK. The equilibrium errors are short memory in respect of Taiwan with Germany, France, and Canada. The null hypothesis of I (0) in the equilibrium cannot be rejected.[11] Future research should explain the economic implication for Taiwan of the long adjustments toward equilibrium with the US and the UK.

Endnotes

1 The I(1) hypothesis cannot be rejected statistically for the individual time series data in Taiwan. See, for example, Wu, J. L. (1995).

2 Cheung and Lai (1993) have shown for any fractionally integrated process that is of order less than one, that the infinite cumulative impulse response equals 0, implying no long-run impact.

3 See Masih and Masih (1995) for whether the condition of $\beta_1=\beta_2$ or $\beta_1=\beta_2=1$ is required or not.

4 See Masih and Masih (1995) for whether the condition of $\beta_1=\beta_2$ or $\beta_1=\beta_2=1$ is required or not.

5 An integral of fractional order is defined as, $C(L)(1-L)^b X_t = D(L)\varepsilon_t$, where, $C(L) = 1 - C_1L - C_2L^2 - \cdots - C_pL^p$, $D(L)=1-D_1L-D_2L^2 - \cdots -$ DqLq , and $C(L)$ and $D(L)$ satisfy the stationary and invertible conditions. We write this ARFIMA (p, d, q) to be
$X_t = (1-L)^{-d}\psi(L)\varepsilon_t = \left(\sum_{j=0}^{\infty} h_j L^j\right)\psi(L)\varepsilon_t$, where $\psi(L) = C^{-1}(L)D(L)$.
When h_0 1,d<1 and $j\to\infty$, $h_j \cong (j+1)^{d-1}$, so we know that any shock to X_t is of hyperbolic decay, or X_t is a long memory process.

6 The advantage of the CSS method over other estimators can be found in Chung and Baillie (1993).

7 The distribution of OLS estimators is now derived by Guo, Lee, Lee, and Wu (1999).

8 The trace test is defined as $LR = -2\ln(Q;H_0 \mid H_1) = -T\sum_{i=r+1}^{n}\ln(1-\hat{\lambda}_i)$.

9 The maximum eigenvalues test is defined as $LR = -2\ln(Q;H_0(r) \mid H_1(r+1)) = -T\ln(1-\hat{\lambda}_{r+1})$, where λ is calculated as the eigenvalues from a matrix derived from a Gaussian VAR model. See Johansen (1991) for details.

10 See appendix for the choice of p and q by AIC criterion.

11 Since the CSS is equivalent to MLE, the t ratio is distributed asymptotically normal.

References

Adler, M., and Lehmann, B. (1983). Deviations from purchasing power parity in the long run, *Journal of Fianace*, 38, 1471-87.

Cheung, Y. W., and Lai, K. S. (1993). Long-run purchasing power parity during the recent float, *Journal of International Economics*, 34, 181-192.

Chung, C. F., and Baillie, R. T. (1993). Small sample bias in conditional sum-of-squares estimators of fractionally integrated ARMA models, *Empirical Economics*, 18, 791-806.

Dickey, D. A., and Fuller, W. A. (1979). Distribution of the estimations for autoregressive time series with a unit root, *Journal of the American Statistics Association*, 74, 427-431.

Diedold, F. X. S. Husted, and Rush, M. (1991). Real exchange rates under the gold standard, *Journal of Political Economy*, 99, 1252-71.

Engel, R. F., and Granger, C. W. J. (1987). Cointegration and error correction: Representation, estimation and testing, *Econometrica*, 55, 251-276.

Frenkel, J. A. (1981). The collapse of purchasing power parities during the 1970s, *European Economic Review*, 16, 145-65.

Gil-Alana, L. A., and Robinson, P. M. (1997). Testing of unit root and other nonstationary hypotheses in macroeconomics time series, *Journal of Econometrics*, 80, 241-268.

Granger, C. W. J. (1986). Developments in the study of cointegrated economic variables, *Oxford Bulletin of Economics and Statistics*, 48, 213-228.

Granger, C. W. J., and Joyeux, R. (1980). An introduction to long memory time series models and fractional differencing, *Journal of Time Series Analysis*, 1, 15-39.

Granger, C. W. J., and Newbold, P. (1974). Spurious regression in econometrics, *Journal of Econometrics*, 2, 111-120.

Guo, M. H., Lee, C. N., Lee, S. Y., and Wu, J. L. (1999). *Asymptotic properties of ordinary least estimators of fractional cointegrating vectors*, Working Paper, Graduate Institute of Economics, National Sun Yat-Sen University, Taiwan.

Hakkio, C. C. (1986). Does the exchange rate follow a random walk? A Monte Carlo study of four tests for a random walk, *Journal of International Money and Finance*, 5, 221-29.

Hosking, J. R. M. (1981). Fractional differencing, *Biometrika*, 68, 165-76.

Johansen, S. (1991). Estimation and hypothesis testing of cointegration vectors in Gaussian vector regression models, *Econometrica*, 59, 1551-80.

Kim, Y. (1990). Purchasing power parity in long run: A cointegration approach, *Journal of Money, Credit, and Banking*, 22, 491-503.

Mark, N. (1990). Real and nominal exchange rates in the long run: An emprical investigation, *Journal of International Economics*, 28, 115-36.

Nelson, C. R., and Plosser, C. I. (1982). Trends and random walks in macroeconomic time series, *Journal of Monetary Economics*, 10, 139-162.

Phillips, P. C. B. (1986). Understanding spurious regressions in econometrics, *Journal of Econometrics*, 33, 311-340.

Phillips, P. C. B., and Perron, P. (1988). Testing for a unit root in time series regression, *Biometrika*, 75, 335-346.

Roll, R. (1979). Violations of purchasing power parity and their implications for efficient international commodity markets. In M. Sarnat and G. Szego (Eds.), *International finance and trends*. Cambridge, MA: Ballinger.

Schwert, G. W. (1987). Effects of model specification on tests for unit roots in macroeconomic data, *Journal of Money Economics*, 20, 73-103.

Sowell, F. B. (1992). Maximum likelihood estimation of stationary univarite fractionally integrated tine series models, *Journal of Econometrics*, 53, 165-188.

Wu, J. L. (1995). Testing long run purchasing power parity in Taiwan, Proceeding in the Open Macroeconomics, *Academia Sinica* (in Chinese), 63-88.

Appendix

Table 13.A1 AIC of estimation fractional integrated process in the PPP equilibrium error

Twn/US

ARFIMA (0,D,Q)	AIC	ARFIMA (1,d,q)	AIC	ARFIMA(2,d,q)	AIC	ARFIMA(3,d,q)	AIC
p=0,q=0	8.06913433	p=1,q=0	8.05933041	p=2,q=0	8.08231631	p=3,q=0	8.07251239
p=0,q=1	8.05933041	p=1,q=1	8.04952649	p=2,q=1	8.06270847	p=3,q=1	8.09661002
p=0,q=2	8.08231631	p=1,q=2	8.07251239	p=2,q=2	8.06270847	p=3,q=2	8.05290455
p=0,q=3	8.07251239	p=1,q=3	8.06270847	p=2,q=3	8.0868061	p=3,q=3	8.04310063

Twn/UK

ARFIMA (0,d,q)	AIC	ARFIMA (1,d,q)	AIC	ARFIMA (2,d,q)	AIC	ARFIMA (3,d,q)	AIC
p=0,q=0	6.63558709	p=1,q=0	6.68113326	p=2,q=0	6.67132934	p=3,q=0	6.66152542
p=0,q=1	6.64128735	p=1,q=1	6.67132934	p=2,q=1	6.67791923	p=3,q=1	6.65172150
p=0,q=2	6.63932661	p=1,q=2	6.66152542	p=2,q=2	6.66811531	p=3,q=2	6.65831139
p=0,q=3	6.63742787	p=1,q=3	6.65988481	p=2,q=3	6.65831139	p=3,q=3	6.64850746

Twn/Fra

ARFIMA (0,d,q)	AIC	ARFIMA (1,d,q)	AIC	ARFIMA (2,d,q)	AIC	ARFIMA (3,d,q)	AIC
p=0,q=0	6.47268362	p=1,q=0	6.49105340	p=2,q=0	6.48630142	p=3,q=0	6.48327719
p=0,q=1	6.46287970	p=1,q=1	6.48630142	p=2,q=1	6.47649750	p=3,q=1	6.48029923
p=0,q=2	6.45307577	p=1,q=2	6.47649705	p=2,q=2	6.47347327	p=3,q=2	6.47049531
p=0,q=3	6.46307448	p=1,q=3	6.48717211	p=2,q=3	6.47049531	p=3,q=3	6.46069139

Twn/Ger

ARFIMA (0,d,q)	AIC	ARFIMA (1,d,q)	AIC	ARFIMA (2,d,q)	AIC	ARFIMA (3,d,q)	AIC
p=0,q=0	6.45326553	p=1,q=0	6.44346161	P=2,q=0	6.47287840	p=3,q=0	6.46307448
p=0,q=1	6.44346161	p=1,q=1	6.47287840	P=2,q=1	6.47249750	p=3,q=1	6.45327056
p=0,q=2	6.43365769	p=1,q=2	6.45327056	P=2,q=2	6.45327056	p=3,q=2	6.45015562
p=0,q=3	6.44327185	p=1,q=3	6.46669358	P=2,q=3	6.44346664	p=3,q=3	6.44708574

Twn/Can

ARFIMA (0,d,q)	AIC	ARFIMA (1,d,q)	AIC	ARFIMA (2,d,q)	AIC	ARFIMA (3,d,q)	AIC
p=0,q=0	7.63192053	p=1,q=0	7.62211661	p=2,q=0	7.61231268	p=3,q=0	7.62356217
p=0,q=1	7.62211661	p=1,q=1	7.63336609	p=2,q=1	7.62356217	p=3,q=1	7.61375825
p=0,q=2	7.61231268	p=1,q=2	7.62356217	p=2,q=2	7.61375825	p=3,q=2	7.60395433
p=0,q=3	7.02508760	p=1,q=3	7.61375825	p=2,q=3	7.60395433	p=3,q=3	7.59415041

14

Re-examination of Monetary Policy in Japan: The Contradiction Between Theory and Practice

Don M. Hong

Professor, Dean, College of Social Sciences and Director, Center for JapanStudies, National Sun Yat-Sen University

Introduction

A government has two arms of macroeconomic policy, fiscal as well as monetary, to ensure economic stability in general. Full employment, stable prices, adequate growth, and balanced international payments can be considered as specific targets to focus on. Fiscal policy is the manipulation of government expenditure and/or taxation. Monetary policy is the action of the central bank to alter the money supply and/or interest rates. In this chapter we discuss only monetary policy to show the conflict between theory and actual operation as exercised in recent times by central banks, in particular the Bank of Japan and the US Federal Reserve. The first section reviews the basic IS-LM macro model in which the money supply is the genuine policy instrument. Then we describe a reality where the interest rate is the policy instrument of choice. In the second section we provide two practical examples of Japan and the US to explain how their central banks drive the economy by steering interest rates, along with an eye kept on the growth rate of the money supply. The final section presents the conclusion.

Theory

The Hicksian IS-LM Model[1] demonstrates that an increase in money supply will cause a shift of the LM curve to the right, and move the equilibrium point E0 to E1, other things remaining the same. The results are an increase in income and a decrease in the interest rate as shown in Figure 14.1. This should be recognized as the consequence of a chain reaction such that (1) the increase in money supply lowers the interest rate along the IS curve which (2) causes an increase in investment, which (3) creates an increase in the level of national income.

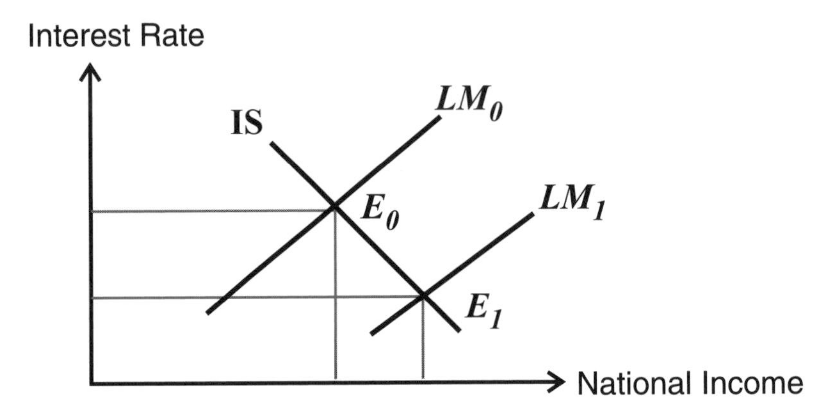

Figure 14.1 IS-LM model

The transmission mechanism through which the money supply influences the physical economy can in turn be further demonstrated in Figure 14.2.

1. Starting point: $M^s = M^d$
2. Then, there is an increase in the money supply: $\Delta M^s + M^s > M^d$
3. Final point: a new equilibrium is established after the demand for money is increased. $\Delta M^s + M^s = M^d + \Delta M^d$

The spillover effect:

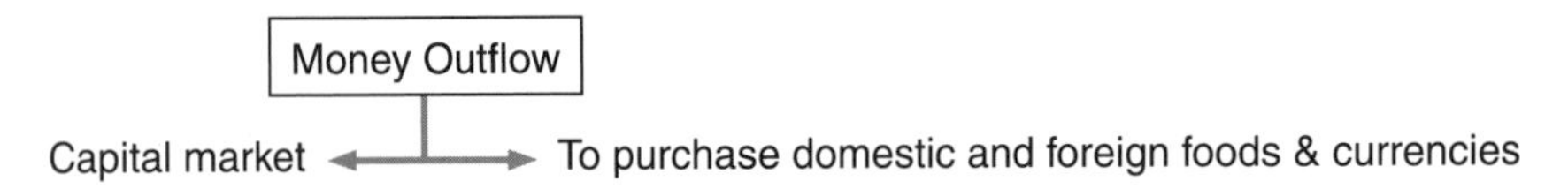

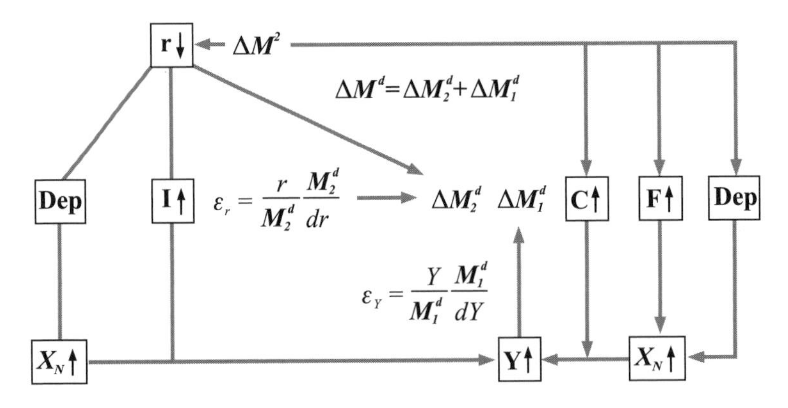

Figure 14.2 The impact of money supply on physical economy

Notations

r	interest rate
I	investment
C	consumption
F	foreign goods and currencies
Y	national income
X_N	net export
M^s	money supply
M^d	money demand
M_1^d	transaction demand for money
M_2^d	speculative demand for money
ε_r or ε_r^M	interest elasticity of demand for money
ε_Y	income elasticity of demand for money
Dep	depreciation of domestic currency
Δ	increment

Interpretation

1. The larger the ε_r (the flatter the LM curve), the quicker the M_2^d being built up, and hence the less the income effect of monetary policy. The argument is consistent with liquidity trap analysis (Keynes, 1936) which points to the ineffectiveness of money on income when ε_r is infinite.
2. The larger the ε_Y (the steeper the LM curve), the quicker the M_1^d being built up, and hence the less the income effect of monetary policy. However, this argument contradicts Friedman's contemplation that "money only matters." Furthermore, among various empirical estimates of the income elasticity of money, Friedman gave us the highest value (1.81) of all (Friedman, 1959).
3. How can both the flatter LM curve (point 1 above) and the steeper LM curve (point 2 above) lead to the same conclusion of a lesser effect of a change in the money supply? By using diagrammatical analysis one can easily resolve the puzzle (Hong, 1979).

In theory, the money supply is exogenous, a variable determined out of the system, while the interest rate is one of the endogenous variables to be determined in the model. The role of the interest rate thus can be thought of as a linkage bridging money, a nominal variable, and production, a real variable. However, from the policy point of view, by controlling the interest rate from the beginning, the economic goal could be more directly and quickly achieved than by controlling the money supply. This can be easily seen from the following formulas first developed by the author:

(1) Money Multiplier:

$$\frac{dy}{dM} = \left(\frac{dr}{dM}\right) \bullet \left(\frac{dI}{dr}\right) \bullet \left(\frac{dY}{dI}\right)$$

(slope of money demand) | *(reciprocal of the slope of MEI)* | *(investment multiplier)*

$$K_M^Y = K_M^r \bullet K_r^I \bullet K_I^Y$$
$$= K_M^r \bullet K_I^Y / K_I^r$$

where K is the money multiplier and MEI is the marginal efficiency of investment.

(2) Elasticity:

$$\frac{M}{Y}\frac{dy}{dM} = \left(\frac{M}{r}\frac{dr}{dM}\right) \bullet \left(\frac{r}{I}\frac{dI}{dr}\right) \bullet \left(\frac{I}{Y}\frac{dY}{dI}\right)$$

(reciprocal of interest elasticity of money demand) | *(interest elasticity of investment demand)* | *(investment multiplier in the form of elasticity)*

$$\varepsilon_M^Y = \varepsilon_M^r \bullet \varepsilon_r^I \bullet \varepsilon_I^Y$$

Graphically (Figure 14.3), the process starts from money and affects investment and consumption via the interest rate and finally national income, which I call a Three-Step Keynesian model.

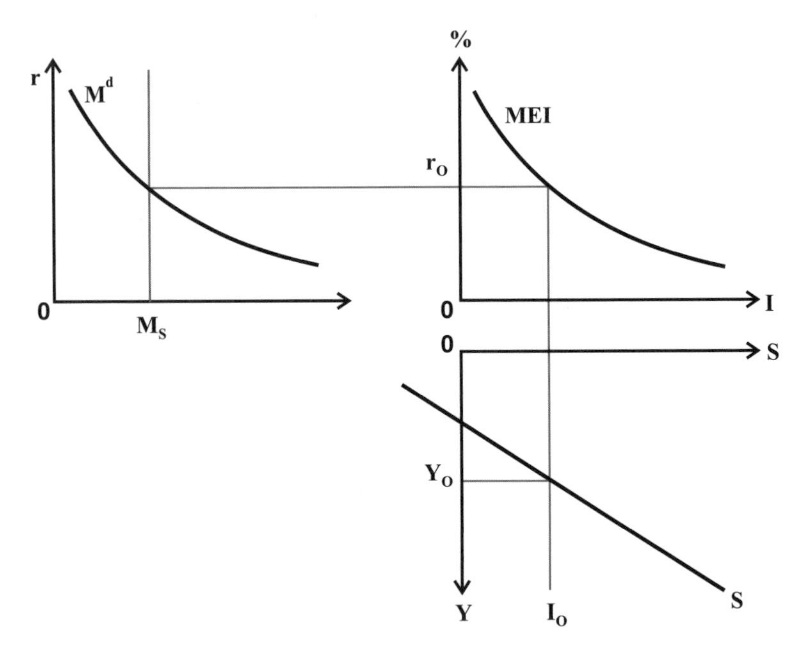

Figure 14.3 Three-step Keynesian model (M→r→I→Y)

This Keynesian idea was first introduced by Samuelson in his renowned economic text book (Samuelson, 1989). In fact, these three steps can be further separated by using the concepts of the multiplier and elasticity, as emphasized above. Due to the fact that the interest rate is considered to be so important in demand functions for both investment and money, we have expanded the naive Keynesian model into the Hicksian IS-LM version, as a step toward reality. However, once the interest rate is predetermined by the monetary authority, investment becomes exogenous as well. As a result, the LM curve becomes horizontal. Hence the equilibrium level of national income will be determined by the equality between the fixed level of investment and a saving function, that is, I=S. Thus the IS-LM model degenerates into the naive Keynesian cross.

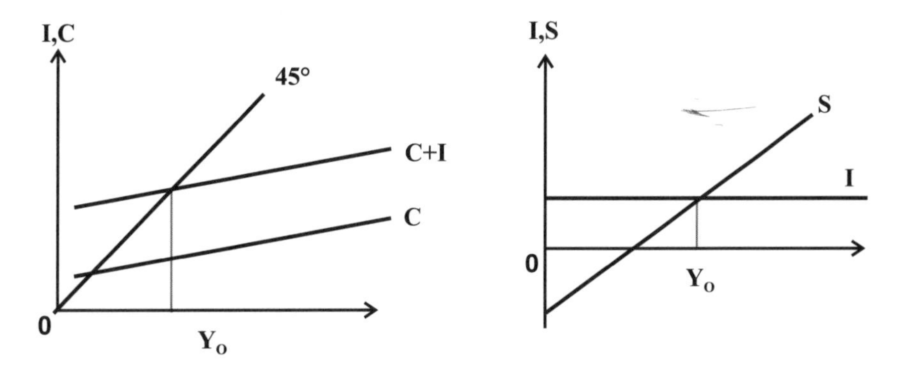

Figure 14.4 Keynesian cross

In the mainstream of economic thought, the Quantity Theory of Money (QTM) dominated macroeconomics before the 1930s. After the Keynesian revolution the sensitivity of money demand to the interest rate was emphasized. However, if M^d (money demand) is unstable and unpredictable, controlling the interest rate is preferable to controlling the money supply as a policy tactic to counter cyclical fluctuations of the national economy. Owing to Friedman's proposal of the policy rule (let the money supply grow annually at the potential growth rate of the economy), the monetary growth rate has been carefully watched by central banks, if it has not been treated as the only tool for conducting monetary policy since the 1970s. Nevertheless, what we have observed in the 1990s is that, while the growth rate of the money supply is considered to be a long term target, the interest rate is utilized as a means of managing the economy in the short run. The impact of a drop in the rate of interest on the economy is further analysed in Figure 14.5 below. The asset effect on consumption is particularly worth noting after a bubble or crash in the stock market.

Practice

Japan is a country that has applied interest rate targeting to an extent hardly matched by other central banks in recent history. First of all, the discount rate was dropped to a historic low at 2.5% in 1987 which triggered the bubble in the late 1980s. Secondly, in 1995, the rate was dropped to 0.5%, a new record low lasting through 1999. Thirdly, in February 1999, Japan forced the overnight call-loan rate between banks downward amazingly to almost zero. As recently as July 1999, for example, the rate was just 0.06% for a one month CD (certificate of deposit), while Canada's rate was 4.72%—80 times higher!

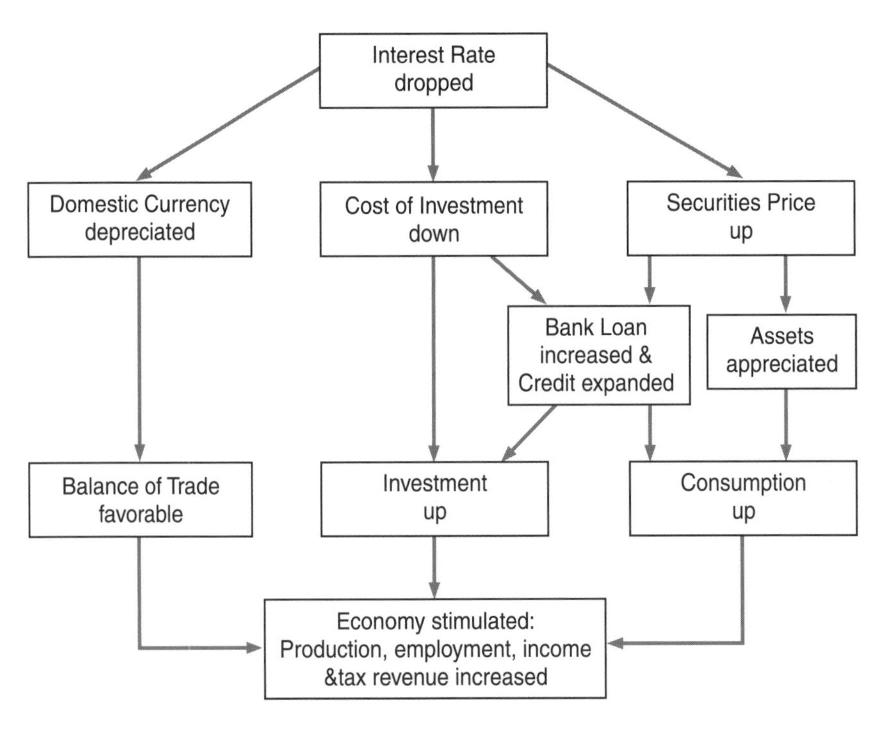

Figure 14.5 The impact of interest on the physical economy

In theory, money supply and the interest rate are inversely related. In reality, however, it may not be so neat as one to one correspondence. Table 14.1 indicates that, between 1995 and 1996, the growth rate of the money supply increased, as did the call rate, but the CD rate decreased; between 1996 and 1997, the money supply increased again, while the CD rate also increased, although the call rate decreased a little; from 1997 to 1998, and again between 1998 and 1999, the money supply increased somewhat; however both the call rate and CD rate dropped greatly.

Money supply growth decelerated from November 1998 (4.3%) through February 1999 (3.4%), and then accelerated from March 1999 (3.7%) through June 1999 (4.3%) before it declined to 3.9% in July and 3.5% in August 1999.

Both year-to-year and month-to-month statistics demonstrate there is a lack of regularity between the money supply and short-term interest rates.

Table 14.1 Japanese financial statistics

Fiscal Year	*Growth Rate of Money Supply*	*Call Rate*	*CD Rate*
1995	2.90	0.49	0.70
1996	3.20	0.72	0.56
1997	3.50	0.70	0.72
1998	3.70	0.05	0.17
1999*	**3.76**	**0.03**	**0.09**
April	4.0	0.03	0.12
May	4.1	0.03	0.07
June	4.3	0.03	0.06
July	3.9	0.03	0.06
August	3.5	0.03	0.06

* Average (April-August)

Source: Japan Economic News, Sept.20,1999. Note: Money Supply=M2+CD

Japan

In the 1980s, "Japan as Number One" was not only a book title (Vogel, 1979) but also a fact believed world wide due to its great economic success. After the 1985 Plaza Accord, the Japanese yen appreciated along with the German mark. The yen, formerly exchanged for the US dollar at 240 yen, appreciated by a staggering 50% in less than 3 years. Because the increased value of the yen failed to reverse the favourable balance of trade in Japan, further US intervention to depreciate the US dollar was expected by the Japanese government. In order to maintain the economy on its high plateau, the Bank of Japan took up a policy to lower interest rates decisively. Beginning from a discount rate of 6.5% in 1985, the rate was eventually dropped to the historic low of 2.5% in 1987 to compensate for the adverse effect of a high yen on export competitiveness as well as to protect against its dampening effect on domestic business conditions through stimulating imports.

The decrease in the interest rate deliberately pursued as a policy during a time when the money supply was increased substantially due to the inflow of foreign reserves created a bubble; not only real estate and securities prices but also the prices of consumer goods and wage rates were raised to record highs. The upshot was that the cost of living in Japan became so expensive that the country was ranked "Number One" in the world.

It is exactly because of this experience that the Bank of Japan failed to increase the money supply directly. Hence, there was no significant trend of monetary growth at a time when the interest rate had a great fall. In fact, the multiplier of money (credit) creation decreased steadily from 13.1 in February 1992 to 10.3 in August 1999, the 20% drop being mainly caused by the weak demand for money (*Japan Economics News*, 20 September 1999). Besides, the actual growth rate of the money supply increased at a rate (4%) greater than potential (3.5%) in the last few years when the economy grew little. This implies that the monetary policy rule could not have worked effectively, even had it been tried. A theoretical dilemma shows up again in implementing monetary policy in Japan.

United States

At the end of June 1999, the US central bank raised its target interest rate on federal funds by 0.25 percentage point to 5% as expected in order to stem inflationary expectations. However, Alan Greenspan, Chairman of the US Federal Reserve, hinted on July 22 and 28, 1999 in Congress that the labour market was very tight, meaning that the wage rate could be higher very soon, and inflation might threaten the economy. This would mean further interest rate hikes.

Early in the autumn of 1998, the Federal Reserve had reduced the rate by 0.75 percentage points in three installments because of the Asian financial crisis spilling over into the US stock market. Since the beginning of 1999, the financial strain has eased and economic activities have firmed in most countries. The world economy has recovered from its 2-year recession and has been regaining strength gradually. In its May 1999 meeting, the Federal Reserve did not increase the interest rate but made a clear announcement that a new policy was adopted that would be leaning in the direction of increased rates, due to a sudden jump in the Consumer Price Index in April by 0.7% per month or 8.4% per year. In the US, the GDP growth rate was as high as 6% in the last quarter of 1998, and 4.3% in the first quarter and 2.3% in the second quarter of 1999. Clearly these actual records (4.2% on the average) overshoot the US potential growth rate (2.5%-3%). Besides, the stock market kept moving upward at a brisk pace in 1999. The Dow Jones Industrial Average passed 11,000 points in May and looked generally steady during the summer of 1999. However, many people feel that the market has long been exuberant and are worried that the bubble could burst unless monetary policy is skillfully managed to create a soft-landing environment. That is why Greenspan has given warnings of an interest rate hike, in order to preempt a sudden collapse in the stock market.

Conclusion

Japan and the United States are the world's two largest economies, but Japan is now suffering from severe deflation while the US has enjoyed prosperity for the past 8 years. To engineer recovery in the Japanese economy, and prevent potential inflation in the US, both countries are focusing on interest rate control as their main policy instrument*. Yet in monetary theory, money supply is the primary means of controlling business conditions. As presented in this chapter, both Japan and the US are behaving in contradiction to monetary theory. On August 24, 1999, the Federal Reserve raised the federal fund rate to 5.25% and the discount rate to 5%. If the Japanese economy fails to recover from deflation, further interest rate hikes in the US might lead to another appreciation of the dollar. As a result, the US trade deficit will continue to worsen, while the Japanese surplus continues to increase, creating an ever-widening gap between the two quantities. In this event, trade conflict between the two countries is inevitable, and protectionism will ensue in place of free trade.

Endnote

1 The IS-LM model was invented by Sir John Hicks and first presented,though in a different form, in Hicks (1937).

References

Friedman, M. (1959). *A program for monetary stability.* New York: Fordham University Press.

Hicks, J. (1937). Mr. Keynes and the 'Classics': A suggested interpretation, *Econometrica*, 5, 147-159 .

Hong, D. M. (1979). *The slope of IS-LM and the effectiveness of fiscal and monetary policy—A graphical exposition*, Taiwan Economy and Finance, Bank of Taiwan.

Japan Economic News, 14.5.98, 18.5.1998, 20.9.1999.

Keynes, J. M. (1936). *The general theory of employment, interest, and money.* London: Macmillian and Co.

Samuelson, P. A. (1989). *Economics*. New York: McGraw-Hill.

Vogel, E. F. (1979). *Japan as Number One: Lessons for America*, Cambridge, Mass.: Harvard University Press.

Wall Street Journal (various).

Plate 16 Yeh Liu, on the northern coast of Taiwan ➧

Symposium Participants

Robert E. Bedeski, Professor, Department of Political Science, University of Victoria

Gerard S.H. Chow, Director, Institute of Mainland China Studies, National Sun Yat-Sen University

Jou Juo Chu, Executive Director, Institute of Public Opinion Research, and Associate Professor, Sun Yat-Sen Institute of Interdisciplinary Studies, National Sun Yat-Sen University

R. Alan Hedley, Professor, Department of Sociology, University of Victoria

Don M. Hong, Dean, College of Social Sciences, and Director, Center for Japan Studies, National Sun Yat-Sen University

Hsin-i Hsiao, Associate Professor, Department of Pacific and Asian Studies, University of Victoria

Ralph W. Huenemann, Professor, Faculty of Business, and Chair of Economic Relations with China, Centre for Asia-Pacific Initiatives, University of Victoria

Michel P. Janisse, Executive Director, Co-operative Education Programs, University of Victoria

Kenneth Keng, Associate Professor and Director, Asia Pacific Research and Development, Faculty of Business, University of Victoria

Richard King, Associate Professor and Chair, Department of Pacific and Asian Studies, University of Victoria

Chingnun Lee, Associate Professor, Institute of Economics, National Sun Yat-Sen University

Stanley T. Lee, Associate Professor, Institute of Interdisciplinary Studies, National Sun Yat-Sen University

S.J. Li, Graduate Institute of Economics, National Sun Yat-Sen University

Ge Lin, Assistant Professor, Department of Geography, University of Victoria

Peter C. Lin, Director, Institute of Economics, and Director, Center of Taiwan Studies, National Sun Yat-Sen University

Tru-Gin Liu, National Sun Yat-Sen University

Warren Magnusson, Professor, Department of Political Science, University of Victoria

Carl A. Mosk, Professor, Department of Economics, University of Victoria

Norman J. Ruff, Associate Professor, Department of Political Science, University of Victoria

John A. Schofield, Professor, Department of Economics and Dean, Faculty of Social Sciences, University of Victoria

Gordon S. Smith, Director, Centre for Global Studies, University of Victoria

David F. Strong, President and Vice-Chancellor, University of Victoria

James H. Tully, Professor and Chair, Department of Political Science, University of Victoria

Marion C.Y Wang, Graduate Institute of Political Science, National Sun Yat-Sen University

S. Anthony Welch, Professor, Department of History in Art, and Executive Director, Office of International Affairs, University of Victoria

Yuen Pau Woo, Director of Research and Analysis, Asia Pacific Foundation of Canada

Yuen-Fong Woon, Professor, Department of Pacific and Asian Studies, University of Victoria

Zheng Wu, Associate Professor, Department of Sociology, University of Victoria